AUTOMATED GUIDED VEHICLES

Dr.-Ing Thomas Müller

IFS (Publications) Ltd, UK
Springer-Verlag, Berlin, Heidelberg, New York
1983

ISBN 0-903608-43-X IFS (Publications) Ltd
ISBN 3-540-12629-5 Springer-Verlag: Berlin, Heidelberg, New York
ISBN 0-387-12629-5 Springer Verlag: New York, Heidelberg, Berlin

©1983 IFS (Publications) Ltd., UK and Springer-Verlag; Berlin, Heidelberg, New York.

The work is protected by copyright. The rights covered by this are reserved, in particular those of translating, reprinting, radio broadcasting, reproduction by photo-mechanical or similar means as well as the storage and evaluation in data processing installations even if only extracts are used. Should individual copies for commercial purposes be made with written consent of the publishers then a remittance shall be given to the publishers in accordance with §54, para 2, of the copyright law. The publishers will provide information on the amount of this remittance.

Publishers:
IFS (Publications) Ltd.
35-39 High Street, Kempston, Bedford, MK42 7BT, UK
and
Springer-Verlag GmbH & Co. KG
Otto-Suhr-Allee 26/28, D-1000 Berlin 10.

Typesetting by Fleetlines Typesetters, Southend-on-Sea, England.
Printed by Anchor Press Ltd., Colchester, England.

Contents

Chapter One	Setting the objectives	3
Chapter Two	What is an AGVS?	5
Chapter Three	Components of an AGVS	27
Chapter Four	Applications for the AGVS	91
Chapter Five	Experience with guided vehicle systems	109
Chapter Six	Planning for AGVS Implementation	115
Chapter Seven	Economic viability as a decision aid in selection	131
Chapter Eight	Reasons for implementing an AGVS	157
Chapter Nine	Development trends	163
Chapter Ten	Summary of AGVS applications	167
Appendix	AGVS installations in Europe	171
References		277
List of illustrations		283

Foreword

SINCE 1960 approximately 220 AGVSs (Automated Guided Vehicle Systems), with a total of approximately 1,300 induction controlled vehicles, have been installed within the Federal Republic of Germany. As well as high shelf storage techniques, considerable development in the automation of internal logistics has been achieved.

This book presents a comprehensive view and opinion of the features, possible applications and economics of the AGVS and is therefore of assistance to planners and operators of transport systems and, in conjunction with robot technology, offers further development possibilities.

Based on experience the author has been able to present problems from within the operational and planning field of AGVSs. This would not have been possible without the assistance of AGVS manufacturers and operators. Special thanks, therefore, are due to the manufacturers of AGVSs and the following companies which have assisted in the economic research.

Bosch-Siemens Hausgeräte GmbH, Berlin;
BMW-AG, Dingolfing;
BMW-AG, München;
Eisenmann KG, Niederlassung Wetzlar;
MAN-AG, Nürnberg;
Mannesmann-Demag, Mannheim;
Siemens AG, Berlin;
Translift GmbH, Grenzach-Whylen;
Thyssen Aufzüge GmbH, Stuttgart;
Vereinigte Verlagsauslieferung Reinhard Mohn OHG, Glütersloh.

Our thanks are due also to I.F.S. (Publications) Limited, for including this work in their programme and for the expedient presentation of this document.

Prof. Dr.-Ing. H. Baumgarten   July 1983

Chapter One

Setting the objectives

AS MECHANISED and automated production plants are expanded the problems of transport, handling and storage become increasingly important, since once these have been overcome there is considerable scope for rationalisation for both industry and commerce (1, p7). The degree of rationalisation which is still available depends amongst other things on the particular branch of industry and the current state of the factory concerned.

Maynard has found, as the result of various investigations in the USA, that on average approximately 30–40% of the manufacturing costs for a product can be ascribed to material flow costs (7, p577). These costs include supply lines to the factory and distribution to the consumers.

In the last two decades the benefits of these cost margins have been gained in several factories through automated storage technology, in particular through high-bay racking technology where the requirements for automated storage could be met, cf. Baumgarten et al. (1). High-bay racking technology, which has now become highly developed, could be quickly implemented and optimised, since in most factories the storage space represents a well-defined area in physical terms with the two interfaces – that is, identification point (entrance) and control point (exit).

This does not apply to internal transport systems which in every respect are more closely tied to the production processes. In this sector each individual load transfer station forms in itself an interface with the adjacent system and in the majority of cases it is considered necessary to have the transport system continuously modernised to keep pace with developments in production technology. This is why lasting success in rationalising internal transport can be obtained only after flexible transport systems with adequate control systems had been developed and which could meet the multiple requirements of materials supply and disposal for manufacturing and commercial enterprises.

The purpose of this book is to examine what applications are possible in today's factories as well as to look at the future for driverless transport systems – or, as they will be referred to here, the automated guided vehicle system – or simply AGVS.

It is almost certain, bearing in mind the trend towards increased automation in manufacturing areas, that the AGVS will occupy an important place in the general field of materials handling.

When planning an automated plant, and hence whether to invest in an AGVS, the first step is to assess the level of investment as well as the expected cost savings. It is also important to establish any other expected benefits which can be measured against other materials handling systems which have already proved themselves in the market place.

Thus, first the various major internal transport systems must be identified and their characteristics classified. Then the various elements of the AGVS must be examined since it is these which determine the technical feasibility of the system. Finally, the areas of application and existing experience of AGVSs must be established in order that the planning engineers are able to provide the necessary basic information.

Once the technical feasibility of a project has been ascertained the economic viability must be studied. This to a large degree determines whether previous planning will be implemented and be successful. In order to limit the effort involved in the technical planning a decision aid should be developed in the early stages which, with the help of cost comparison curves based on a computer program, give information on whether the planned handling system alternatives are tenable from the point of view of operating costs and whether they still can be incorporated in future planning. The reasons for installing AGVS and further development trends will determine the scope of future applications of driverless transport systems.

The organisation, structure and contents of this book are designed to meet these objectives. In addition, a study of the application of driverless transport systems is complementary to a large extent with the issues which arise when planning such systems.

This study therefore is intended also to be of use to planners and users of transport systems whose function Gudehus (3, p145) has defined in the following manner:

'A transport system should be structured, dimensioned and organised so that a defined transport operation is performed optimally with respect to the given technical, spatial and temporal boundary conditions.'

Chapter Two

What is an AGVS?

WITHIN THE FRAMEWORK of this study internal transport systems will be defined as all trucks that are used for internal handling of the material flow throughout industrial and commercial enterprises. Internal transport mainly has the task of bridging horizontal distances and handles material flow inside buildings, but it also includes transport between buildings provided that the public transport system is not involved. According to this definition, lifting machinery, cranes and elevators as well as trucks where horizontal and vertical transport is of more or less equal importance, such as stacking cranes, are excluded from considerations of internal transport systems.

When mainly horizontal tasks are involved continuous and discontinuous transporters (truck systems) are used, cf. Figs. 1 and 2. Following (11) continuous transporters also include power-and-free conveyors since they are more closely related to chain conveyors than to trolley conveyor systems.

Fig. 2 shows the driverless transport system or AGVS with respect to other driverless overhead transport systems (monorail conveyors) and conventional floor-mounted and driver-operated transport systems. Overhead and floor-mounted hybrid systems are formed when certain functional elements of the monorail conveyor, namely the bearing structure, the drive system, the power supply and the steering are partly free of the floor and partly mounted on the floor. Thus, for instance, with the scooter the loading units are transported on a floor truck with its own drive while the power supply and the control of the steering system are provided by an overhead rail as with the monorail conveyor (42).

Hybrid systems are also produced when rail-bound trucks running along the floor can be lifted by elevators into overhead bearing structures (Autover ^{R}).

The concept of an automated guided vehicle system embraces all tran-

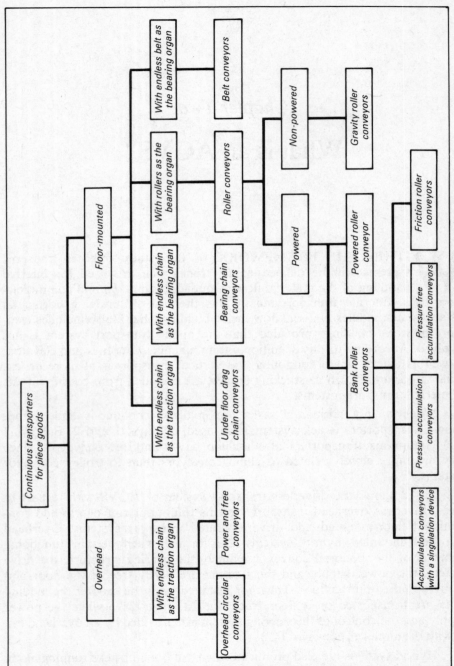

Fig. 1: Classification of continuous transporters for piece goods

WHAT IS AN AGVS?

sport systems which are capable of functioning without driver operation. As the result of the development of inductive steering (see chapter 3) it was first possible in 1954 to use driverless battery-driven electric tractors in the USA with inductive steering (43).

At the same time the concept of 'driverless' came into being applied in particular to those industrial trucks which would conventionally be driver-operated. Meanwhile the concept of driverless transport systems (DTS) or AGVS has passed into general use for describing battery-driven industrial trucks with contactless steering (17,36).

The main components of the AGVS are:
- ☐ The truck or tractor, pallet truck, tow skid basic type;
- ☐ The floor system with the installation of the wire guidance system and the information transfer system;
- ☐ The load transfer equipment which can be both on board the truck and/or in a stationary position, including the station structure;
- ☐ The truck and traffic control system.

A detailed description of all components of the system is given in Chapter 3.

The AGVS is an especially flexible system for horizontal transport and is suitable for both simple transport operations with a small number of destinations as well as complex and centrally-controlled transport processes. The flexibility of the system lies in the relatively problem-free laying of the guidance wire which is supplied with high-frequency current responsible for the inductive steering of the trucks.

The economic laying of the guide wire makes it possible to build up internal transport network which can be used for nearly any kind of transport operation provided that there is enough space available. Only when there is a limited number of destinations and a very high traffic density are the continuous transporters clearly superior to the AGVS (see Chapter 7).

The VDI recommendation 3562 classifies truck systems using mechanical steerling in the same group of driverless industrial trucks, together with trucks with inductive steering, provided they are battery-driven. These systems however have not gained very much practical importance. Instead, mechanically steered systems have partly come into use when the power is also provided by a contact line laid along the floor (UTS [R] and System Mobil [R]), (Fig. 2.).

Whereas with inductive steering an electromagnetic field is produced by a cable in the floor and is picked up by coils in the truck for steering, when the trucks are steered by sensors the active and passive elements are swopped over — that is, the control sensors on the truck are guided by a passive control line (44). Trucks with sensor steering have not gained any importance in industry and commerce; the same applies to systems with optical steering which because of their susceptibility to dirt and the relatively heavy demands on the floor could not win any acceptance, (Fig. 2).

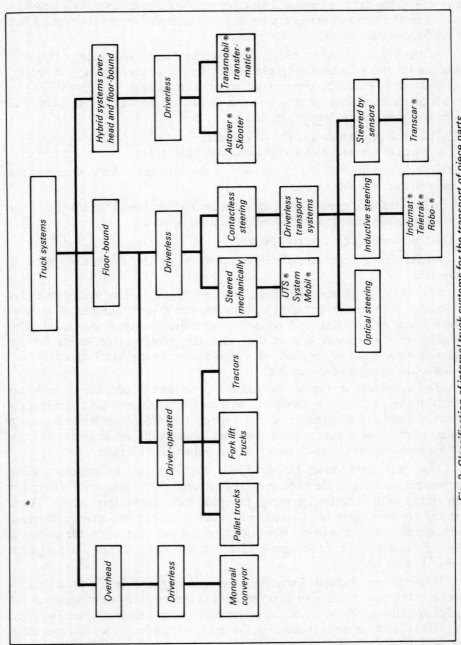

Fig. 2: Classification of internal truck systems for the transport of piece parts

WHAT IS AN AGVS?

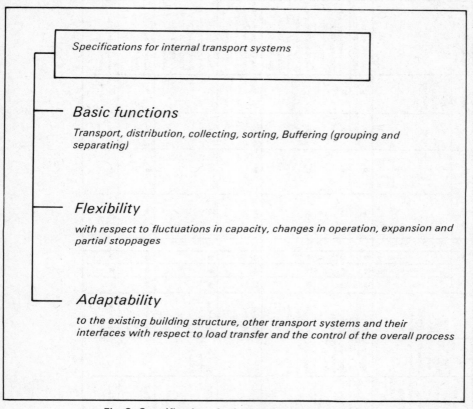

Fig. 3: Specifications for internal transport systems

Characteristics. When first introduced on the market the AGVS was in competition only with driver-operated systems (forklift trucks, electric tractors and related industrial trucks). As a general rule systems with 1-10 AGVS trucks were used.

The use of on-board microprocessors on the truck and the possibility of process control by a central computer have, in conjunction with a mature truck technology, led to a situation where complex driverless systems are also now in competition with other transport systems as listed in Figs. 1 and 2 (e.g. monorail conveyor and roller conveyor systems).

The current market situation for complex transport systems is characterised by demanding requirements which are shown in general terms in Fig. 3, and by corresponding further developments in the transport systems which are presented in Figs. 4 and 5 with their basic characteristics.

These characteristics include:

☐ The conveying technique; continuous or discontinuous as well as overhead or floor-mounted,

AUTOMATED GUIDED VEHICLES

| No. | Means of transport | Conveying technique | | | Conveying speed in m/sec | Transport path in m | Conveying capacity per truck in loading units/hour | Theoretical conveying capacity of the system in loading units/hour | Type of drive/ energy consumption | Goods being transported/ handling aid | Performance | | | | Flexibility with respect to | | Adaptability | | Load transfer |
		Continuous	Discontinuous	Overhead	Floor-bound							Transporting	Distribution	Collection	Buffering	Fluctuations in traffic density	Organisational changes in material flow	Existing building structure	Other transport systems	
1	Hand pallet truck (Driver-operated truck systems)		x		x	1	< 50	approx. 20	—	Manual	Mainly pallets	◐	◐	◐	○	◐	●	●	◐	Manual/active conveyance
2	Fork lift truck (Driver-operated truck systems)		x		x	1.4–2	< 200	approx. 15–30	—	Electric motor battery	Mainly pallets	◐	◐	◐	○	◐	●	●	●	Mechanical/active conveyance
3	Tractor (with 3 trailers) (Driver-operated truck systems)		x		x	2–3.3	≥ 100	for 100 m approx. 60	—	Electric motor battery	Palletised/unpalletised goods	●	◐	◐	○	◐	●	●	◐	No independent load transfer possible: conveyance passive possibly station active
4	Tractor (with 3 trailers) (Driverless transport systems (DTS) with inductive steering)		x		x	$V_=$ = 1 V_i = 0.35	≥ 100	for 100 m approx. 35	400		Palletised/unpalletised goods	●	●	●	○	◐	●	◐	◐	Fully automatic load transfer possible: chain conveyors or roller conveyor
5	Pallet trucks (Driverless transport systems (DTS) with inductive steering)		x		x	$V_=$ = 1 V_i = 0.35	≥ 100	for 100 m approx. 11	250	Electric motor Battery	Palletised goods	◐	●	●	◐	◐	●	◐	◐	Automatic load handling: only possible by backwards travel, automatic load delivery
6	Lifting platform trucks (Driverless transport systems (DTS) with inductive steering)		x		x	$V_=$ = 1 V_i = 0.35	≥ 100	for 100 m approx. 18	250		Palletised/unpalletised goods	◐	●	●	○	◐	●	◐	◐	Automated load transfer
7	Designed as a stacker (Driverless transport systems (DTS) with inductive steering)		x		x	$V_=$ = 1 V_i = 0.35	≥ 100	—	250		Palletised goods	◐	●	●	○	◐	●	◐	●	Automated load transfer

WHAT IS AN AGVS?

| No. | Means of transport | Conveying technique | | | | Conveying speed in m/sec | Transport path in m | Conveying capacity per truck in loading units/hour | Theoretical conveying capacity of the system in loading units/hour | Type of drive/ energy consumption | Goods being transported/ handling aid | Performance | | | | Flexibility with respect to | | Adaptability | | Load transfer |
		Continuous	Discontinuous	Overhead	Floor-bound							Transporting	Distribution	Collection	Buffering	Fluctuations in traffic density	Organisational changes in material flow	Existing building structure	Other transport systems	
8	Monorail conveyor		x	x		0.33	≥ 100	—	250	Electric motor via contact line	Depends on the individual load handling aid	◑	●	●	◐	◐	◐	◐	●	Automated
9	UTS® System Mobil®		x	x		1	≥ 100	—	250		Palletised/ unpalletised goods	◐	◐	◐	◐	◐	◐	◐	◐	Automated load transfer possible truck active, station active
10	Autover®		x		x	4	—	300		Electric linear motor	Palletised goods	◐	●	●	◐	◐	◐	◐	◐	Automated with telescopic forks
11	Transfermatic®		x		x	—	—	Depending on the assembly process		Electric motor via contact line	Assembly workpieces up to medium weights	◐	●	●	◐	◐	◐	◐	◐	Automated
12	Overhead conveyor	x		x		0.3	≥ 100	400			Dependent on the individual handling unit	●	◐	●	◐	◐	◐	◐	◐	Automated load transfer possible
13	Power and free	x		x		0.36	≥ 100	400		Electric motor. Chain as traction unit		●	◐	◐	●	◐	◐	◐	◐	
14	Underfloor drag chain conveyor	x			x	0.3	≥ 200	200			Palletised/ unpalletised goods	●	◐	◐	◐	◐	◐	◐	◐	Automated load transfer only possible under certain conditions

Rows 8–11: Truck systems with mechanical steering
Rows 12–14: Continuous transporters with chains as the traction units

No.	Means of transport	Conveying technique				Conveying speed in m/sec	Transport path in m	Conveying capacity per truck in loading units/hour	Type of drive/ energy consumption	Goods being transported/ handling aid	Performance				Flexibility with respect to		Adaptability		Load transfer
		Continuous	Discontinuous	Overhead	Floor-bound						Transporting	Distribution	Collection	Buffering	Fluctuations in traffic density	Organisational changes in material flow	Existing building structure	Other transport systems	
Continuous transporters – roller conveyors																			
15	Gravity roller conveyor	x			x	—	—	—	Gravity	Palletised/ unpalletised goods, boxes	●	◐	◐	●	◐	●	◐	◐	Stripper possible
16	Powered roller conveyor	x			x	0.3	> 50	500	Electric motor	Palletised/ unpalletised goods, containers	●	◐	◐	◐	◐	◐	◐	◐	Automatic load transfer possible
17	Accumulation roller conveyor	x			x	0.3	> 50	500	Sliding clutch gears		◐	◐	◐	●	◐	◐	◐	◐	
18	Friction roller conveyor	x			x	≤ 0.2	> 30	—	Electric motor. Friction gears	Assembly workpieces up to heavy weights	●	◐	◐	◐	◐	◐	◐	●	
Other transport systems																			
19	Chain conveyor	x			x	0.3	—	500	Electric motor	Pallets, skeleton boxes, containers	◐	◐	◐	◐	◐	◐	◐	●	Automatic load transfer possible
20	Trolleys	x			x	< 0.15	—	—		Assembly workpieces	◐	◐	◐	◐	◐	◐	◐	◐	
21	Belt conveyor	x			x	< 1	—	1000		Containers cardboard boxes	●	◐	◐	◐	◐	◐	◐	◐	

● Very well suited
◕ Well suited
◐ Suitable
◑ Relatively suitable
○ Unsuitable

V_v Forwards speed
V_r Backwards speed

Fig. 4 and 5: System overview of the main characteristics of internal transport systems

WHAT IS AN AGVS?

- ☐ The conveying speed,
- ☐ The economic transport path (average values from experience),
- ☐ The conveying capacity,
- ☐ The type of drive,
- ☐ The goods being transported or handling aids,
- ☐ The degree of fulfilment of the specifications defined in Fig. 3.

A further additional characteristic which needs to be borne in mind is the degree of automation associated with the load transfer at the interfaces and the overall process control.

In principle all transport systems can be provided with information which makes it possible to automate the material flow of the system. However, it must not be forgotten that the only functions which can be automated are those which a transport system performs using its special technical facilities (e.g. collection, sorting, buffering, load transfer). To this extent the possible degree of automation can, in respect of the achievable performance, be put down to the basic functions which the transport system is technically capable of carrying out.

Discontinuous transporters. In many enterprises internal transport is performed by forklift trucks, especially if a flexible and manoeuvrable industrial truck is required which is capable of picking up the load by itself and, should the situation arise, be able to perform stacking operations if necessary.

For mixed use both inside and outside large buildings, cantilever four-wheel trucks come into use. They can pick up loads outside the wheelbase and they possess a high stability. Large wheel diameters facilitate good acceleration.

If the gangways are in good condition the three-wheel high-lift truck can be used also. In contrast to the four-wheel truck it is usually driven from the rear axle (one wheel) and so cannot achieve its driving performance. However, it is characterised by its better manoeuvrability and it is more inexpensive as a result of its simpler design. Almost without exception an electric motor drive is used with a battery as the power source for transport indoors; occasionally internal combustion engine driven trucks with fuel gas as the power source are used.

Battery-driven reachmast trucks, such as those in Fig. 6a, combine the advantages of cantilever and wheel-supported trucks (extending trucks). The load is picked up in the same way as with cantilever trucks, and they can pick up the load on ramps with the reach mast extended. If the reach mast is retracted into the wheelbase a favourable load distribution for horizontal travel is obtained; at the same time the working aisle width is reduced in storage areas (18).

Forklift trucks are universal industrial vehicles with a broad range of applications which partly overlap those of the AGVS. It makes sense for AGVS to replace forklift truck operations when the forklift truck is used mainly as a multipurpose industrial truck or is mainly or solely used for long horizontal operations, as happens in practice time and time again (38).

With sufficient traffic density such transport operations can be performed more economically by dividing up the operations so that the forklift trucks carry out the loading and unloading, for example from AGVS tractor trains, while the AGVS takes over the actual transport operations. Furthermore there are transport operations where the DTS carries out the entire transport including load transfer by itself by means of its own load handling equipment or with the help of load transfer machines, see chapter 3.

Electric tow tractors or electric lift trucks with seated drivers who can steer are typical of the industrial trucks which are exclusively used for horizontal transport, see Figs. 6b and c.

Since with forklift trucks a large part of the investment costs are associated with the lifting frame, electric tractors and related types of industrial trucks are considerably more economical.

As far as applications are concerned such systems are in direct competi-

Fig. 6: Driver-operated industrial trucks with electric motor drive (battery). (a) Reach mast truck with driver and three-wheel construction. (b) Pallet truck with driver. (c) Truck tractor with four-wheel construction

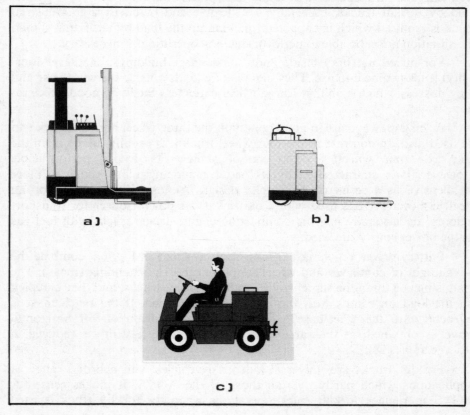

WHAT IS AN AGVS?

tion with the AGVS and they will always have an economic advantage over the AGVS when sporadic transport operations have to be carried out over extensive and constantly changing transport paths.

On the other hand for transport operations with a material flow density which justifies using several trucks and an organisational structure which can be mapped by a transport network the AGVS is usually superior to the electric tractors and electric pallet trucks, especially when production-integrated operations have to be carried out at the same time which are dangerous, for example when driving through automatic palletising machines and shrink-wrapping ovens.

Details of the economic applications of these trucks can be found in chapter 7.

Continuous floor transporters. Roller conveyors and chain conveyors, Fig. 1, are continuous conveyors and suitable for transport, collection and distribution. Often roller conveyors and chain conveyors are used at the interface between a storage area and the internal transport system as decoupling units. They serve as buffers as well as interface transport systems which provide the transfer to the load handling units of the storage trucks (for example, stacking cranes) and to the load handling units of the transport system (for example, AGVS). Fig. 7 shows the components of a roller conveyor system.

Accumulation roller conveyors occupy a special position among continuous transporters because the task of filling in time by buffering is of the same importance as other functions. Accumulation conveyors are always used in a conveying system when the delivery and disposal of transport units cannot be synchronised.

Typical applications are:
- Accumulation sections ahead of merging paths in continuous transporter systems.
- Accumulation sections ahead of intermittent conveyors such as vertical conveyors, sidetracking trucks, turn-tables and swivel tables.
- Accumulation sections ahead of junctions and transfer equipment such as roller lifting tables and chain transfer devices.
- Accumulation sections ahead of production and processing machines, palletising and depalletising machines as well as packaging machines.
- Storage sections as intermediate buffers and terminating sections.
- Accumulation conveyors in live storage systems.

Accumulation conveyors can be divided into three basic technical types in accordance with Fig. 1:
- Accumulation conveyors with a single drive.
- Pressure accumulation conveyors.
- Pressure-free accumulation conveyors (10).

Accumulation conveyors with single drive are the most expensive

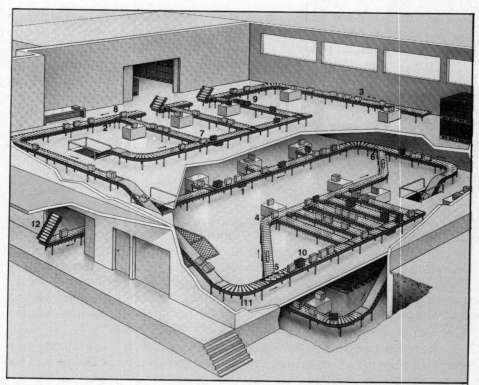

1 Drive
Powers up to 100 m sections

2 Drive shaft
Transmits the drive over straight sections, bends and branches

3 Roller conveyor – straight section

4 Roller conveyor bend
Transports round bends of 30, 45, 60 and 90° and sums of these angles

5 Outward points 30°
Moves the goods from main sections to branch sections

6 Inward points 30°
If there are gaps transfers goods from branch sections to main sections

7 Outward transfer device 90°
Stops the goods and transfers them to the branch section

8 Inward points 90°
If restricted space replaces inward points 30°

9 Accumulation section with stop device or rocker gate
Stops the goods. Controls the frequency on sections

10 Pressure-free accumulation section
Stops each item individually. When the front accumulation position becomes free it moves each item one place forward

11 Supporting frame
Supports the roller conveyor at the required operating height

12 Non-powered section
Transports on straight sections and bends using gravity

Fig. 7: Components of a roller conveyor system including belt conveyor for overcoming gradients

solution to the problem of buffering, both in terms of technology and the investment. This is because a single drive unit with its electrical control is necessary for each accumulation position. Accumulation conveyors with single drive are used when exact positioning is required at the accumulation point.

Pressure accumulation conveyors have the following characteristics in common:

☐ The transport units are piled up without any space between them.

WHAT IS AN AGVS?

- [ ] On accumulation a pressure is produced between the transport units.
- [ ] The support rollers are driven with slip.

With the pressure-free accumulation conveyors the transport units are piled up without any pressure with spaces between them. In order to achieve this there are sections on the conveyor with a defined number of positions. The drive of these sections is engaged and disengaged by control vanes or sprockets using a mechanical lever system.

A description of the various control systems used for pressure-free accumulation conveyors can be found in (11).

Overhead circular conveyors with chain drive. Overhead circular conveyors with chain drive (power-and-free conveyors) consist of the basic components: track, tension chain, drive station, tensioning station, carriage and suspension gear with load supports.

Overhead mechanical circular conveyors with chain drive transport and suspended unit loads are inevitable. The transport path can be relatively freely adapted to the spatial conditions providing the overall suspended load (chain conveyor + transport load) can be supported by the existing building structure.

Additional structures for supporting the chain conveyor are usually bound up with considerable costs; therefore under certain circumstances it is impossible to use chain conveyors economically. If one ignores the permissible forces which can be passed into the building structure then the carrying capacity of a chain conveyor depends on the loading capacity of the carriages and of the track (9). Standard loads are less than 200kg per carriage. If several carriages are involved, (as for car body plants) the carrying load can be increased up to 1000kg.

The track lengths are usually 100–500m so that large systems must be made up of several chain conveyors. High utilisation of the load-bearing capacity of a chain conveyor gives favourable dead load: payload ratios which on average result in a value of approximately 1:10.

One version of the overhead chain conveyor is the circular traction conveyor, generally known as the power-and-free conveyor. Power-and-free conveyors have the facility for removing the transported goods from a section of the conveyor, stacking them, storing them or transferring them to another conveyor section (9).

The power-and-free conveyor consists of a traction system and a loading system, see Fig. 8. The traction system consists of the track, the traction unit and the running gear. In principle it is constructed in the same way as the conveying system of a circular conveyor but without suspension gear and load carriers.

Traction running gear differs from circular conveyor gear by having dogs which are mounted. The loading system consists of the loading track and

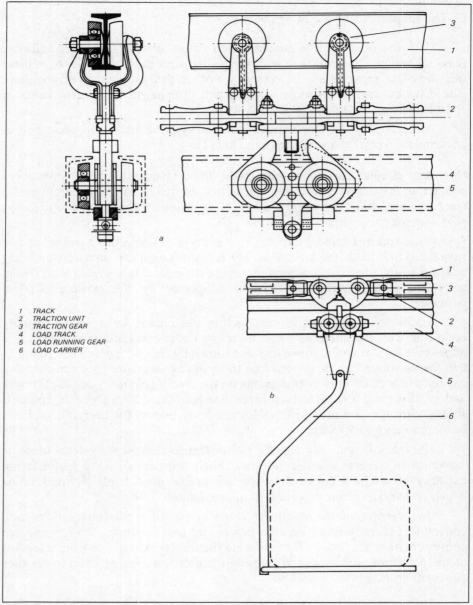

Fig. 8: Components of the circular traction conveyor (power and free). (a) Traction unit beneath the traction track. (b) Traction unit inside the traction track

1 TRACK
2 TRACTION UNIT
3 TRACTION GEAR
4 LOAD TRACK
5 LOAD RUNNING GEAR
6 LOAD CARRIER

loading gears, suspensions and load carriers; it can therefore be compared with a trolley conveyor without travelling gear (19).

The loading system is capable of working together with several traction systems so it is possible in one system to work with different conveyor speeds

WHAT IS AN AGVS?

using transport with and without mechanical operation and with any branching of the conveying path. Each traction system in such a system is closed in and has its own drive. The loading system must be designed so that access is ensured to the main section at all times so facilitating the circular movement of the load carriers.

With very simple systems the control of the destination is carried out by an accompanying address on the trolley by means of cams (reference value) and by interrogating switches at key points on the track. The control of the destination routing is usually made by employing the latest technology using an electronic comparison of actual and reference values without any contact being made.

The throughput in loading units per hour is a function of the speed of the chain conveyor and the distance between the individual load carriers (that is, load carrier spacing). With an average speed of v = 0.3 m/s and a load carrier spacing of approximately 3m for small unit loads equation (1) gives an hourly throughput of 360 loading units provided that all load carriers are utilised.

$$(1) \quad Q_{P+F} = \frac{3600 \cdot v}{t_L} \text{ LU/h (number of loading units per hour)}$$

where Q_{P+F} = the throughput of the power-and-free conveyor in loading units per hour
v = the chain speed in m/s
t_L = the load carrier spacing in m

The advantage of the overhead chain conveyor is its high and constant throughput. At the same time production-related processes (for example, immersion baths for surface treatment and assembly processes) can be integrated.

With the help of loops dynamic stores can be generated to provide time buffers between individual manufacturing processes. In addition power-and-free conveyors provide the facility of disengaging the load carriers from the traction path and so can perform sorting operations; in this case the individual sorting buffers must not be permanently in circulation as with proper circular chain conveyors. Thus the power-and-free conveyor provides a much more flexible system for solving transport problems.

However, compared with such monorail conveyors as overhead systems and floorbound driverless transport systems there are some disadvantages in the operation of chain conveyors. With power-and-free conveyors the relatively high noise level is especially noticeable because of the engaging and disengaging of the load carriers. Relative to the monorail conveyors and the AGVS the chain conveyors are at a disadvantage with respect to

☐ Speed of the transported loading units.
☐ Flexibility of the travelling paths.
☐ Possibilities of destination control systems.

This is because these systems, with independent individual drive for each truck, can cope with a large number of material flow inputs and outputs in any combination. The speeds of 1–1.5 m/s are approximately four times higher than that of the chain conveyors.

The choice of a suitable transport system will be determined finally by the specifications which can be derived from the transport objectives. For high throughput with relatively fixed transport operations it makes sense to use chain conveyors whereas the monorail conveyors and the AGVS have the advantage when it comes to complex conveying paths, noise development and the possibilities of destination control. The flexibility of the truck systems (monorail conveyors and AGVS) will come into its own in terms of conveying path design when it becomes necessary to expand the transport network.

With chain conveyors the conveying path is a dynamic element by virtue of the endless chain which is designed with a certain length and for reasons of construction cannot be extended arbitrarily. With the truck systems the transport network can in principle be extended indefinitely.

Overhead conveyors with separate drives. The monorail conveyor is a discontinuous transporter and belongs to the truck systems shown in Fig. 2. It is used mainly for parcel or piece goods transport operations. Bulk materials can be transported if they are presented as part or parcel loads in containers.

The monorail conveyor consists on the one hand of the driving mechanism and on the other hand of the rail system with the directional components such as points and lifting and lowering stations. The main components of the monorail conveyor are shown in Fig. 9.

Fig. 9: Components of the monorail conveyor

1	MOTOR
2	DRIVE WHEEL
3	TRAILER CARRIAGE
4	LINK
5	CONDUCTOR RAIL
6	CURRENT COLLECTOR
7	OVER-RUNNING BRAKE
8	MOUNTING PLATE
9	LOAD CARRIER
10	RUNNING RAIL

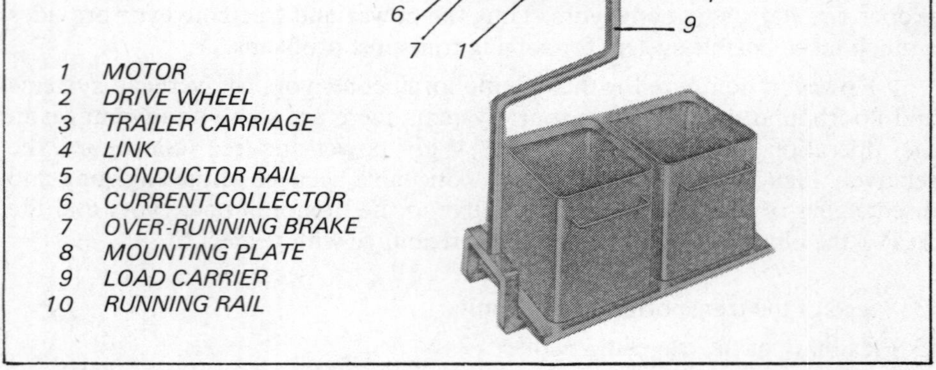

WHAT IS AN AGVS?

The moving gear consists of a drive unit and occasionally a trailer carriage which has an articulated connection with the drive unit. Usually each individual truck carries one loading unit with the help of a runner which may be equipped with gripping and holding devices. However, there is also the possibility of pulling one or several trailer carriages with a guide mechanism (a so-called tractor). Aluminium sections are used predominantly as the running rails but special requirements must be made of the running surface at high speeds (> 0.5 m/s).

In principle, monorail conveyors can be used when the material flow density does not justify the use of a continuous transporter and when a floor-bound conveying technique is preferable (20). Within a rail system monorail conveyors are useful also for irregular transport operations, such as when the transport frequency on certain sections can be controlled according to demand by moving trucks out of a station (truck buffer).

As a result of the independent drive of each truck any destination can be linked with any other within a rail network. This is why the monorail conveyor is a system which like the AGVS can meet the high requirements of destination control facilities.

In medium-size networks with increasing material flow density the monorail conveyor comes into competition with the power-and-free systems, especially when the buffering function also has to be fulfilled as in Fig. 10.

With medium traffic density in conjunction with a high degree of automation in the destination control and load handling systems the monorail conveyor system comes into competition with the driverless transport systems. Therefore when assessing the performance capacity of the monorail conveyor within the framework of systems planning the power-and-free conveyors and driverless transport systems must also be taken into consideration at the same time.

The length of monorail conveyors is a function of the material flow and the functions of the system. With most of the installed monorail conveyor systems the conveying length lies in the range 250–2,500m. Occasionally there are cases where the conveying route is as much as 4,000m.

The theoretical maximum transport performance of a suspended rail system is determined by the possible truck sequences in the rail system. It is a function of the travelling speed in m/s and of the distance between the individual trucks in metres. The distance between the individual trucks is a function of the length of the truck or the length of the load and the braking deceleration in m/s^2, as shown in equation 2 below:

(2) $X_A = \dfrac{v^2}{2b} + L_F$ m

where X_A = the distance between individual trucks in m
  b = the braking deceleration in m/s^2
  L_F = the length of the individual truck or of the suspended load unit in m

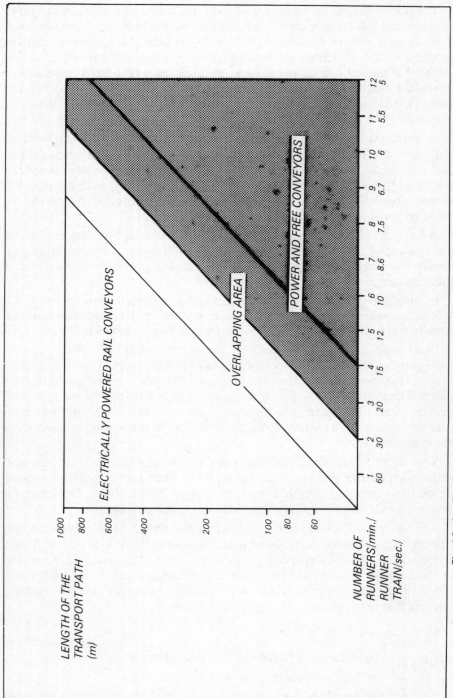

Fig. 10: Application areas of electrically powered rail conveyors and power and free conveyors

v = speed in m/s

In practice, this theoretically possible performance of an electric suspended rail conveyor is considerably reduced by the times which have to be added on as the result of turning points, lifting and lowering operations and queuing at junctions. Moreover, the time necessary for load handling which lengthens the running time of a transport operation must also be considered.

The standard truck speed for the monorail conveyor is approximately 0.5m/s. On fast straight sections speeds of up to 3.3m/s can be obtained. In this case multi-speed motors or variable mechanical gears are used.

When negotiating bends the speed should be matched to the radius of the bend and to the goods being transported. The same applies when entering points and lifting and lowering stations.

With a truck length L_F of 1m, an average speed of 0.5m/s and a braking deceleration of $0.3 m/s^2$ the distance between the trucks comes to $X_A = 1.42m$ according to equation 2.

The transport capacity of the system is then calculated from equation 3.

(3) $Q = \dfrac{v}{X_A}$ (trucks/s)

where Q = the transport capacity of the system (trucks/second)
 v = the speed in m/s
 X_A = the distance between the trucks in m

Using the figures quoted the equation gives a Q of 0.35 (trucks/sec). This example then gives a theoretical hourly capacity of 1,270 trucks or loading units.

In practice, however, the distance between the trucks is increased by the block control system which keeps the trucks at a specified distance from one another during operation. In principle, block control systems function in the same way as with driverless transport systems. In this connection see also Chapter 3.

In order to keep the costs of block control systems within acceptable limits a distance is selected between the individual trucks which is larger than the theoretically necessary distance X_A. In connection with waiting times and other switching and non-productive times hourly transport capacities are obtained which lie far below the theoretically determined value of the above example. With a sufficient number of available trucks transport capacities of 250–300 (loading units per hour) have been obtained in practice.

Other assembly and handling systems. In the manufacture of components and final assembly there are two basic forms of transport systems.

The first of these is a transport system using classical 'point to point' technology whereby a certain sequence of operations at different work places are linked together continuously, in-phase or out-of-phase. The main disadvantage of this technology is that the configuration is restricted from the point of view of production control because the result of a complete working

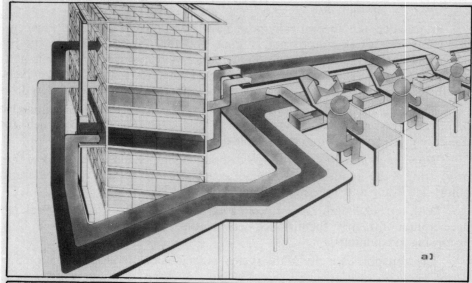
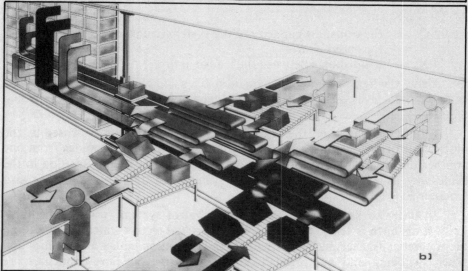

Fig. 11: Handling systems for servicing work places: (a) Container conveyor systems in conjunction with a stacker crane-operated live store; (b) Stacker crane system for servicing the work places and the pigeon-hole store (dual operation)

sequence with respect to the criteria of time-limits (work progress), quantity control, and quality control is usually only visible at the end of a working sequence.

Typical conveying systems belonging to this production technology are assembly lines, platform conveyors, and conventional industrial trucks such as forklift trucks and so on. Although work progress can be predetermined with the help of mechanical transport on assembly lines the disadvantages are that the planning of the assembly parts requirements must be very accurate and there is increasing resistance to assembly line work on the part of the personnel.

The second form of basic transport system is based on the principle that the necessary material is supplied from a central storage area for each operation and that this is returned to the store on completion of this operation. In other words each operation is individually serviced in terms of supplying and removing material.

With the return of material to the store every operation can be centrally booked and quality control and quantity control can be carried out at the same time. Through central stocking and handling the storage of materials at the workplace is minimised and there is therefore no need for direct synchronisation between the work places.

There is considerable transparency in production control through the availability of information from the central store. Furthermore the material flow can be directly tied to the production schedule, that is the sequence of operations is predetermined by the material flow.

Such systems, also called 'central work distribution' (55), have gained considerable importance in the clothing and electrical industries (45). In the main container conveyor systems are used as transport to the work place together with pigeon-hole stores or live stores. The material flow in the store can be handled with the help of a stacker crane as in Fig. 11a. However the work places can also be serviced directly with a stacker crane even at low material flow densities, cf. Fig. 11b.

In certain applications (depending on the industry) there is no advantage in storing the material centrally after each operation. The reasons for this are the effort involved in transporting large work pieces and a quality control which can only be effective at the end of the entire sequence of operations (for example, the assembly of engines in the car industry).

Under these conditions transport systems present themselves which operate with automatic work-piece carriers and so also eliminate the requirement for synchronisation. Furthermore, they enable random access to the work piece carriers by means of points and buffer sections, (23). The disadvantage of these systems replacing assembly lines lies in their higher investment costs and higher space requirements.

In principle the requirement for synchronisation can be eliminated or, in other words, 'central work distribution' can be implemented with the AGVS or "UTS"[R], as described in chapter 2. In some cases the monorail conveyor can also be used.

Examples of applications for the AGVS are also given in Chapter 4.

Chapter Three

Components of an AGVS

DRIVERLESS transport systems consist basically of trucks and a traffic system. From the physical and information point of view five basic components can be defined and which correspond to one another.

On the physical level these are the truck, the network and the load handling system; on the information level these are the truck control and the traffic control systems, as shown in Fig. 12 below.

Fig. 12: Overview of the most important system components of driverless transport systems

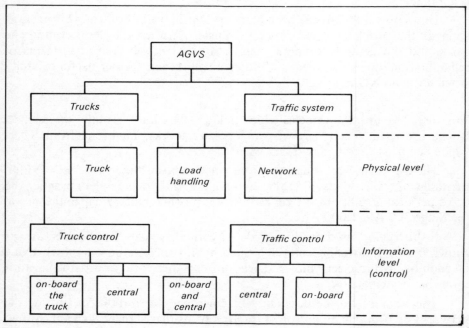

The load handling system represents a common component belonging to the truck side and the network side on the physical level. The following three possibilities have to be considered.

- [ ] the truck is exclusively active
- [ ] a stationary load handling station is exclusively active
- [ ] the truck and the load handling system are both active.

A distinction must be made between truck control and traffic control on the information level.

Apart from other functions the truck control system is concerned mainly with destination control. In the first instance traffic control carries out the job of preventing collisions on sections and at junctions.

In most cases truck control and traffic control are carried out on board the trucks. However, in complex systems it makes sense to control the trucks from a central computer especially when the driverless transport system communicates with several interfaces (such as load transfer to chain conveyors) which in their turn have to be controlled.

In these complex cases the central computer utilises the truck's onboard logic; that is it can send the command 'drive from A to B' to the truck while the truck carries out the actual destination control itself.

In special cases this function also can be performed centrally so that the control system is completely centralised. Since the information transfer between trucks, network and central computer is totally dominated by induction systems an additional installation expenditure is involved if this solution is adopted.

Up to now little use has been made of central traffic control. However, as soon as the problem of information transfer between the network and the truck and the central computer has been satisfactorily solved in terms of reliability and costs this means of control will release further useful capacity when using an AGVS.

Tractors. Tractors are vehicles which pull a rolling load. Usually three-wheel trucks are used which, as swivelling bolster trucks, have one front wheel which can be steered and two rear wheels.

The drive can be through the front wheels and/or the rear wheels. Within the range of trailer loads of 2,000–20,00kg currently occurring in practice the drive performance lies in the range 0.75–7kW while battery capacities are in the range 300 Ah–1,000 Ah.

If the truck is used partly outside a hybrid drive can then be used as was shown in the Hanover fair in 1982. Such hybrid vehicles have a diesel engine in addition to the electromotor drive and can therefore be provided with a lower battery capacity.

The tractors can be designed (for emergency operations or outside the network) as pedestrian-controlled tractors or as driver-seated tractor. Some-

COMPONENTS OF AN AGVS

times the driver can stand at the steering controls. The maximum speed is 1m per second which must be reduced on bends and gradients.

Only in areas completely well protected from pedestrian and industrial truck traffic are higher speeds permissible in accordance with the safety regulations.

The truck's brakes generally are designed as electromagnetic disc units. In the event of an emergency stop the drive motor can be operated also as a generator. For reason of cost industrial trucks involved in inside operations should take up as little room as possible. Because of this the steering behaviour of a truck during cornering plays a special role in terms of space requirements.

The steering behaviour of a truck on cornering is determined by the following key factors:

☐ the steering system comprising swivelling bolster steering, axle pivot steering, differential steering and all-wheel steering.
☐ the distances between axles and in relation to this
☐ the truck dimensions in the horizontal projection.

Given the same maximum lock of an axle pivot steering system the swivelling bolster steering shows the same behaviour when cornering. This is characterised by the tail of the truck running inwards so that it is not true-tracking, taking more room in the inside curve than for example an all-wheel steered truck (see Fig. 13).

Since there are no special stability requirements when the tractor is used inside a factory, the swivelling bolster steering system in the three-wheel design has become more popular than the axle pivot steering system because it is simpler in construction and it allows a steering lock up to 90°. In this case the truck can turn about point P, as shown in Fig. 13.

When an entire truck train is involved in automated load handling then the envelope curve of the entire train should be closely followed when cornering with swivelling bolster steering in order to leave enough clearance in the inside curve on the one hand and to align the truck train after the curve as quickly as possible if there is to be a load transfer with a defined interface.

The behaviour of the tractor on bends can be improved by completing the cornering manoeuvre with a compensating movement as in Fig. 14. In addition true-tracking trailers can be used in long tractor trains.

There are different ways of transferring the load to and from tractor trains. First of all we have to differentiate between tractor trains which always behave as a closed unit and those tractor trains where the trailers can be coupled and decoupled according to demand.

Tractor trains that function as a unit are mainly used in cases where there are relatively few destinations and there is a large material flow. In this case load transfer is often effected by stationary units such as loaders and

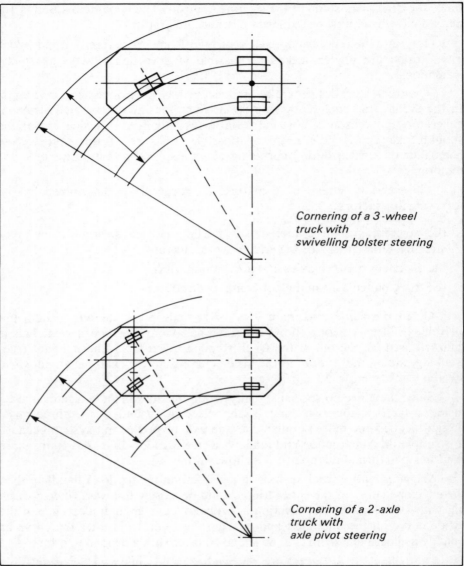

Fig. 13: Steering systems for industrial trucks, swivelling bolster steering and axle pivot steering systems

telescopic fork devices. The load transfer is often also effected using fork lift trucks.

Load transfer stations will be understood as all stationary equipment from a dumping point for loading units right up to a fully automatic loading station. If there are very few destinations the lowest investment costs for load handling are obtained when the station is exclusively active during load handling operations. In this way load handling can be combined with other

COMPONENTS OF AN AGVS   31

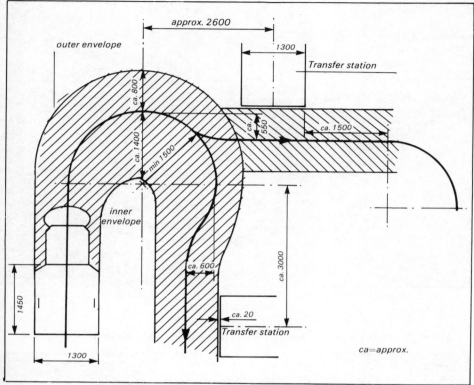

Fig. 14: Envelopes for a truck with fifth-wheel steering

production operations, for example a palletiser integrated in the loading operation.

If the number of tractor trains is roughly the same or less than the number of destinations then the investment costs for load handling are in general minimised by having the tractor train exclusively active during load handling; for example by means of trailers with lift platforms which remove the pallets from jibs.

When deciding which is the best load handling system to use the constraints of the interfaces must also be taken into consideration. Under certain circumstances it may make sense to have both the tractor train and the load transfer station active during load handling.

What is mainly involved here is the transfer of pallets from stationary conveyors to conveyor segments mounted on the tractor train (chain conveyor and roller conveyor). But tractor trains with lifting platforms are also used to lift the pallets from stationary chain conveyors or deposit pallets on to them. Figs. 15–19 show applications where the tractor train functions as a closed unit.

If trailers are coupled and decoupled during transport operations this can be carried out either manually or automatically. If automatic coupling is used

Fig. 15: Load transfer between a tractor train with lifting platform trailers and a stationary chain conveyor

Fig. 16: Load transfer from a tractor train to a conveyor by means of telescopic forks

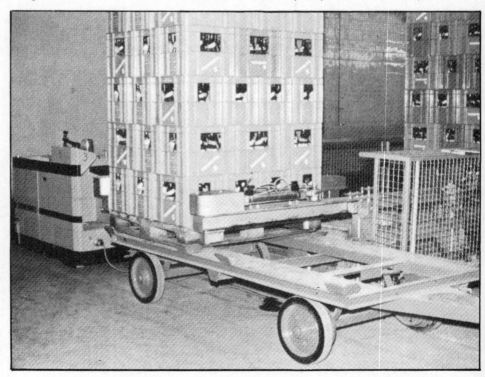

COMPONENTS OF AN AGVS

Fig. 17: Loading of a tractor train inside a palletiser

Fig. 18: Sequential unloading at the interface to a high-bay racking store with subsequent contour testing. Truck and station are equipped with chain conveyors

Fig. 19: Simultaneous load transfer between stationary roller conveyors and trailers with roller conveyor segments mounted on them

COMPONENTS OF AN AGVS

the tractor must be able to reverse; a self-engaging bolt coupling is sufficient for this. If automatic coupling and decoupling is required the coupling can be performed electromagnetically provided that the trailer loads are not too large.

When the trailer is decoupled the load transfer has usually not been completed; that is, the load has still to be taken from the trailer or brought to its exact destination on the trailer. Forklift trucks are especially useful here and, if necessary can be equipped with true-tracking construction as well as trailers with block and steering rollers.

Figs. 20, 21 and 22 show examples of applications for tractor trains with different methods of coupling. In particular, Fig. 20 shows the bolt coupling for automatic coupling. The arm with the induction search coils of the magnetic field can be clearly seen below the truck. The exposed control unit is still fitted with conventional relay technology.

Fig. 21 shows an electromagnetic coupling and Fig. 22 a tractor train with true-tracking pallet trucks which can be individually positioned at the place of application.

A semi-trailer truck is shown in Fig. 23 as a special version of the tractor train as used in the area of the lowest rack level as might be found in the food wholesale trade.

When tractors are used particular attention must be paid to inclines. If any short-term loading on the electric motors is based on the VDE regulations then gradients of approximately 3% can be negotiated with standard dimensions provided that the adhesion value μ of the floor is at least 0.5.

Fig. 20: Reversing tractor with bolt coupling

AUTOMATED GUIDED VEHICLES

Fig. 21: Reversing tractor with electromagnetic coupling for automatic coupling and decoupling operations

If steeper and longer gradients have to be negotiated under load then the manufacturer must make special allowances in the design in terms of the drive motors and battery capacities. For down gradients which are negotiated under load or which also have bends the trailers must be equipped with brakes.

Pallet trucks. Inductively steered pallet trucks as a general rule transport one loading unit. In special cases pallet trucks with fork lengths for 2 loading units (pallets) are also used. The payloads lie in the range 1,500–2,000kg; the drive

Fig. 22: Tractor train with true-tracking pallet trucks as trailers

COMPONENTS OF AN AGVS

Fig. 23: Semi-trailer truck with roller containers in commissioner applications for wholesale operations

performance is approximately 0.7kW and the battery capacity 210–400Ah depending on the conditions.

Speeds of 1m per second are achieved on straight sections and on bends the speed is reduced to 0.5m per second.

As with most tractors the pallet trucks are designed with three-point linkage. The front wheel is constructed as a driven fifth wheel and the steering behaviour as well as other features correspond to that of the tractor.

The trucks are supplied either with exclusively forwards automatic movement or automatically forwards and reversing movement.

With forward moving trucks the load transfer is usually carried out manually, driving the truck under the loading unit using the drawbar. Automatic unloading is carried out by lowering the forks at the destination. Automatic loading of forward-moving trucks can be achieved by means of an additional load transfer station with lifting and lowering equipment.

The inductively steered pallet truck with automatic reversing has an extra pair of induction coils underneath the end of the forks at the tail end, moving at a speed of 0.25m per second under the loading unit. Compared to the tractor this truck is especially suitable for applications with a low material flow density and a relatively large number of destinations at which the load can be transferred exclusively through the activity of the truck.

Typical of the standard loading units which can be transported with the pallet truck are the Europallets, box pallets and purpose-built roller pallets or roller containers. If the goods to be transported consist of small containers

Fig. 24: Inductively steered pallet truck; drawbar towing

which can be manually loaded and unloaded the loading surface can be in the form of a platform. The lifting and lowering equipment with its own drive and hydraulic pump can then be dispensed with. In mixed applications trucks are also used where the forks can be covered with an upward folding platform, as shown in Figs. 24, 25 and 26.

A broad range of applications of individual transport operations can be performed with the help of pallet trucks, especially of the automatic reversing type, because the trucks handle the load transfer themselves at the destination without any stationary load transfer equipment being necessary and with full automation of transport and load transfer the desired rationalisation is achieved.

Fig. 25: Inductively steered pallet truck with fork length for 2 pallets and infra-red distance control, cf. chapter 3.5.3.2

COMPONENTS OF AN AGVS

Fig. 26: Inductively steered pallet truck with upward folding platform

Fig. 27: Unloading of load units (forwards travel) using pallet trucks

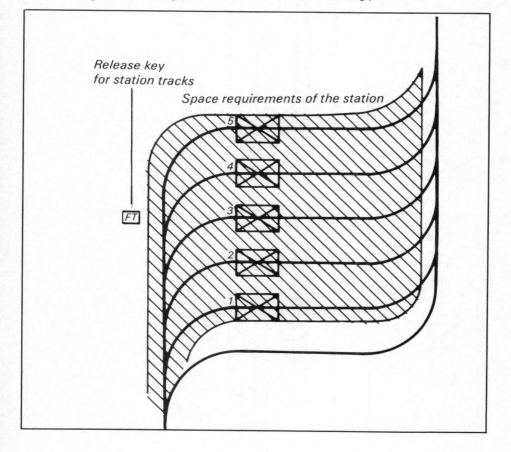

Due to the truck characteristics applications involving inclines, and elevating sections, can be carried out more simply than for example in the case of tractors. However, it should not be forgotten that with a large number of destinations and a relatively high traffic density the space requirements of pallet trucks become a crucial factor.

Fig. 27 shows the amount of space required by a forward moving truck at an unloading point for one pallet and for several pallets.

Loads can also be set down by reversing. In this case the pallets are placed at a distance from each other of approximately 150mm with the help of an infra-red sensor at the end of the forks. When picking up loads the truck control system is informed by a special cap on the back of the forks when the forks are completely underneath the pallet, Fig. 28.

Fig. 28: Load handling (unloading and/or picking up) with automatically reversing pallet trucks

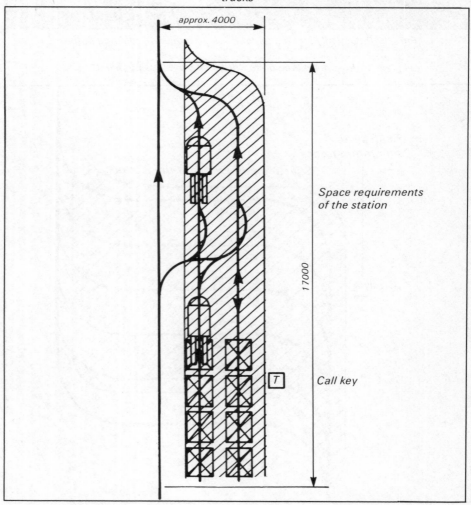

COMPONENTS OF AN AGVS 41

Fig. 29: Inductively steered pallet trucks in commissioner applications with lifting equipment

If automatically reversing trucks are used additional space must also be provided for reversing travel. In this case an additional consideration is that usually as a result of reversing (crawling speed) a section on the main thoroughfare has to be closed to traffic. This produces an additional relatively long time to be added on to the time of the load transfer operation and this can have an effect on the number of trucks which have to be used. In this connection see also (21).

Pallet trucks are used also in commissioner applications; two basic types of truck are used here:

- ☐ The standard pallet truck as shown in Fig. 24; in this case the frequency of approach within one rack row is relatively high and the equipment is started and stopped by a commissioner working next to the truck. The commissioning is kept entirely to the bottom rack level.
- ☐ Pallet trucks with a pedestrian driver steering; in this case the frequency of approach within a rack row is lower and the commissioner goes from one rack unit to the next on the truck itself. The trucks can be equipped with lifting equipment so that the commissioner and the loading unit can be lifted up to the second rack level, see Fig. 29.

Both types of truck can employ flat pallets with lattice mounting frames as well as roller containers depending on the conditions.

Skid tractors. Skid tractors are trucks which do not pick up the loads at floor level. The load is either taken from a console, a conveyor or a loading unit is used which has enough clear height. The lifting and lowering equipment on the truck is usually active during load handling, as shown in Figs. 30, 31 and 32.

In certain cases the load can be lowered on to the truck with a lifting jack. Alternatively the truck tows a roller container as a trailer loading unit with the help of a retaining bolt; when this towing principle is used active load handling (lifting and lowering equipment) is not necessary on board the truck.

Skid tractors are highly manoeuvrable trucks since the truck length, apart from the safety hooks, essentially corresponds to the length of the loading unit, as highlighted in Figs. 31 and 32.

The trucks are usually designed according to the principle of differential steering or all-wheel steering which provides the additional advantages:

- ☐ forward and reverse travel can be performed at the same speed;
- ☐ minimum space is required when cornering with given truck geometry. Compared to fifth-wheel steering and axle pivot steering (Fig. 13) differential steering and all-wheel steering requires a smaller path width, that is, $X_3, X_4, < X_1, X_2$ (Fig. 33); with $X_3 = X_4$.

Differentially steered trucks have two drive wheels situated in the middle of the trucks sides; these are responsible for steering.

COMPONENTS OF AN AGVS

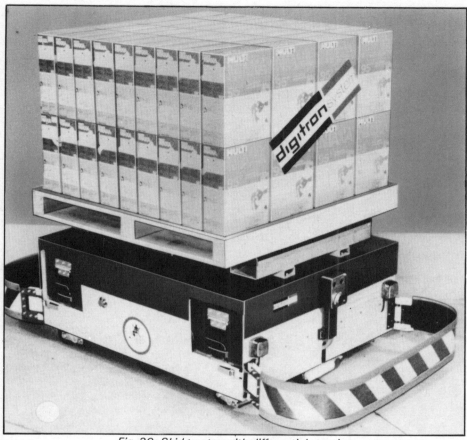

Fig. 30: Skid tractor with differential steering

Fig. 31: Skid tractor transporting a motor including the loading unit

Fig. 32: Skid tractor transporting a raised roller container

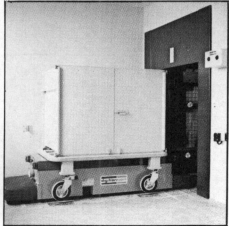

Cornering is achieved by driving the outer wheel at a higher speed than the inner wheel. Due to the completely symmetric construction of the truck, (Figs. 32 and 33) any direction of travel (e.g. forwards travel) can only be defined on the basis of the last reversal of direction; this makes differentially steered trucks especially suited for shuttle operation in blind alleys.

All-wheel steered trucks present favourable steering behaviour similar to differentially steered trucks. Insofar as inductively steered trucks are equipped

Fig. 33: Steering systems of industrial trucks, differential steering and all-wheel steering

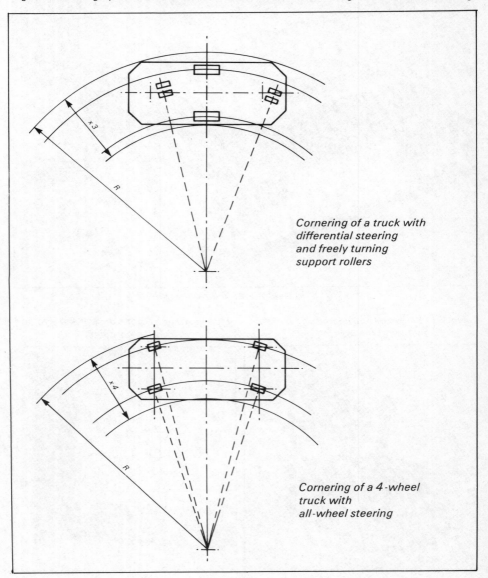

Cornering of a truck with differential steering and freely turning support rollers

Cornering of a 4-wheel truck with all-wheel steering

COMPONENTS OF AN AGVS

with all-wheel steering usually only the wheels on the left or righthand side of the truck are controlled, that is the wheels on one side of the truck turn freely as idlers.

Since only the three-wheel base is defined statically on the ground it can happen that the truck rests on the two idler wheels and only on one steered drive wheel. This produces an undefined steering condition. In order to overcome this problem the truck frame is designed along its length in such a way as to permit torsions. In conjunction with a suitable elasticity characteristic of the idler rollers a four-point base is obtained with stabilised steering behaviour (39).

The all-wheel steering system is preferred for large trucks since the manoeuvring capacity is further improved through the possibility of transverse travel, especially during load handling, (Fig. 34). In principle diagonal travel is also possible. Fig. 35 shows an all-wheel steered truck in production-integrated operation with a lifting platform.

Skid tractors are very versatile trucks which can be fitted in many different ways to provide applications as transport platforms, often integrated into the production process. The payloads can vary from 300 to 3000kg; truck lengths vary from approximately 2m to approximately 5m; other features of these trucks are identical to those of the trucks previously described.

Fig. 34: Steering systems of industrial trucks, transverse travel and diagonal travel with all-wheel steering

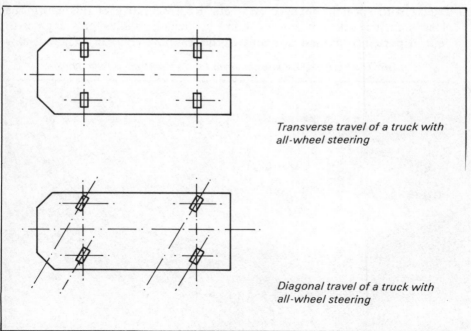

Transverse travel of a truck with all-wheel steering

Diagonal travel of a truck with all-wheel steering

Fig. 35: Skid tractor as a transport platform in production-integrated operation (welding line)

Special trucks. Based on the truck types already presented a large number of other types have been developed for special uses. Three main development trends are involved here.

☐ Trucks with special conveyor elements. Representative of this enormous class is a truck with a heavy link conveyor which enables lorry engines to be transported from final assembly to the test bays (Fig. 36). There is also

Fig. 36: Truck with heavy link conveyor for transporting lorry engines

Fig. 37: Truck with roller conveyor for handling loads during commissioning

a truck with a chain conveyor which carries out the material flow from a high-bay store to the commissioner points and is characterised by its manoeuvrability and short length of 2.0m including the safety hooks (Fig. 37). The comparable length of pallet trucks is approximately 2.7m.

☐ Trucks with reach forks or telescopic forks outside the wheel base. With these trucks load handling can also be carried out without a load transfer station being involved. Also limited stacking operations can be performed at the same time as a result of the load being picked up outside the wheel base.

Fig. 38 shows a reach mast truck which operates according to the principle of the side loading truck; it can stack loading units approximately 1m high in two tiers.

Fig. 39 shows a reach mast truck transporting three roller containers at the end of a packaging line for milk products and storing these roller containers in a cold store provided with a two-tier rack. Unloading on to the bottom level takes place at floor level.

Fig. 40 shows a truck with a lift mast and two telescopic forks for

Fig. 38: Reach mast stacker in the side-loading version

carrying two containers. Through the transport operations of the truck production line work places are serviced from a pigeonhole racking system with three tiers.

□ Trucks in production-integrated assembly. In production-orientated assembly the trucks are usually adapted to the special conditions of their application both in terms of their loading gear and their construction. Although these trucks have not been able to replace classical transport systems (conveyors) in the large-scale manufacturing sector they have proved themselves in the small-lot production sector as far as the following criteria are concerned, namely flexibility, elimination of the need for synchronisation during production, and reliability. Production-integrated trucks occupy an important position in the market for driverless transport

Fig. 39: Reach mast stackers transporting three roller containers and handling loads in a cold store with rack levels

Fig. 40: Servicing of production work places with the help of an inductively steered telescopic fork lift truck

systems (see also chapter 9). Although the number of installations is relatively small the number of trucks per plant lies in the range 10–360 with several plants having 100 and more trucks. In this respect special constructions can be justified, such as those shown in Fig. 41 for the final assembly of car engines. Fig. 42 shows a truck which is used in the aggregate assembly of chassis.

Apart from the truck types already mentioned inductive steering is gaining importance in other areas of industrial truck applications. Thus use is made of inductive steering in the automation of stacker cranes, (40). Moreover efforts are being made to apply inductive steering to reach mast stackers in block stores. The first results on the introduction of these trucks on the market are given in (47).

We shall not concern ourselves further with these applications and trucks

Fig. 42: Special truck in the aggregate assembly of chassis, car industry

Fig. 41: Special truck used in the final assembly of car engines
Dead weight   300 kg
Payload   300 kg

and the associated trucks since such inductively steered trucks are being used in the first instance to service storage areas and not to provide internal transport facilities.

The selection of the type of load transfer system considerably narrows down the choice of the type of truck; at the same time the way the whole system is controlled is affected also.

For this reason the definition of a suitable load transfer system is of central importance in planning a driverless transport system especially when a large number of load transfer points have to be reached with an extensive system.

The first decision to be made is whether the load is to be handled at floor level or free of the floor as a function of the transport system and the existing or planned internal interfaces. It then has to be decided whether during load transfer

☐ only the truck
☐ only the load transfer station
☐ or the load transfer station together with the truck is active.

The determining factors for this decision are the number of load transfer points in relation to the number of trucks, with consideration given to the economic viability of the process as well as the required throughput in load units per time unit given that queues can build up at bottlenecks (e.g. at the load transfer point at a high-bay store).

COMPONENTS OF AN AGVS 51

With a large number of load transfer points in relation to the number of trucks load transfer exclusively carried out by the truck will be usually the most economic solution. With material flow densities as can occur at the input end and during accumulation at the output end the throughput performance may have to be improved by an active load transfer station. Furthermore the internal restrictions, such as the existing loading appliances, interfaces to existing conveying systems and the amount of available space have to be taken into consideration.

The determining factors for load transfer with driverless transport systems are given in Fig. 43.

With large complex systems the behaviour of the sub-systems and the variations in loading of the systems over a given time also have to be considered in relation to load transfer. This means, for example, that in the high-bay store the driverless transport system has to be uncoupled at its interface using adequate buffering as the result of different and stochastic cycle times of the individual systems.

This applies especially when one sub-system first has to be stepped up as the result of the loading process, such as peak activity at goods-in. Routing is especially important here; it is common to use bypass tracks at the central load transfer point of the high-bay store. Chapter 4 deals with the best way of designing the high-bay store/AGVS interface.

Based on the approach in (52) a systematic classification of load transfer possibilities is shown in Fig. 44. Apart from the distinction made beween load acceptance (loading) and load discharge (unloading) the 'active' and 'passive' states of the load transfer station and truck are distinguished. The types of trucks and load transfer stations are classified accordingly.

This classification reveals a large number of load transfer opportunities resulting from the combination of different loading and unloading possibilities with varying degree of automation. The best solution for load transfer in any particular case is given by the cost-effectiveness ratio which can be evaluated with the help of the structure given in Fig. 43 for any application.

Truck control. Most on-board truck control systems have the following functions:

☐ Steering control which keeps the truck on its course

☐ Route control which brings the truck through the network to the predetermined destination. This function can be performed also by a central computer outside the truck.

☐ Drive control which first starts up the truck motor and then accelerates, brakes and stops the vehicle as a function of the traffic situation.

☐ Load appliance control which in the case of a mechanised load appliance controls loading and unloading on the truck (this function can also be performed by a central computer).

In order to be able to implement these control functions there must be a

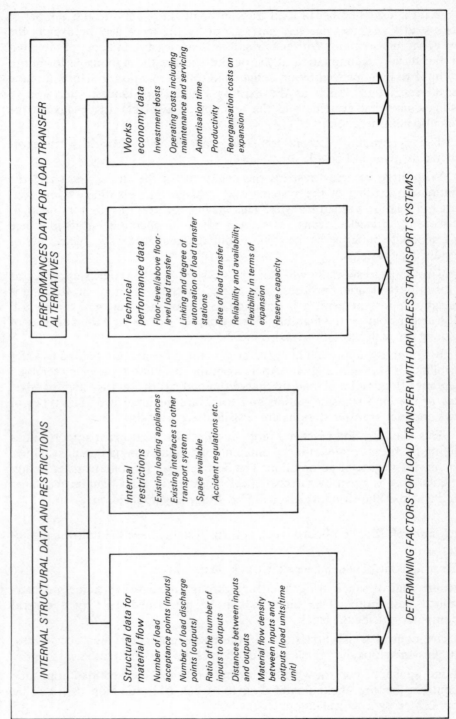

Fig. 43: Determining factors for load transfer with driverless transport systems

COMPONENTS OF AN AGVS

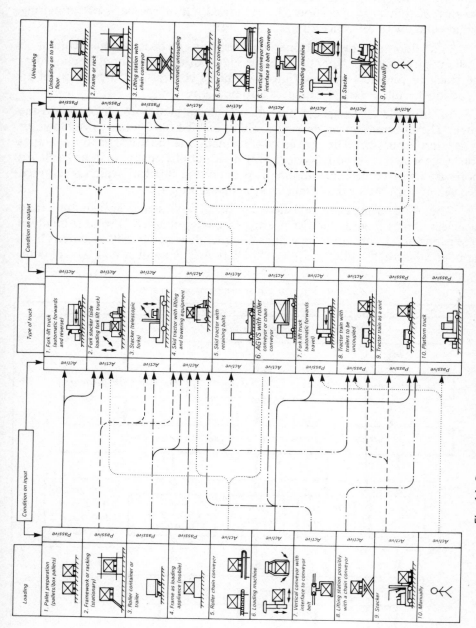

Fig. 44: Systematic classification of load transfer for driverless transport systems

means of transferring data between the truck and the floor installation so that the current transport operation can be carried out on the basis of comparing actual values with reference values.

In addition the transport operation itself has to be issued as a command to the control system. This takes place either through an input keyboard at the truck or remotely by input into the truck's control logic by means of an induction circuit. Combined input is also possible.

In principle all transport operations can be incorporated on board the truck. The information processing is either carried out by hardwired logic or by a microprocessor. The processed information is then passed to the actuation stage which is responsible for carrying it out physically. Insofar as closed loops are involved new information is produced in this way which is returned to truck control by the data transfer system, see Fig. 45.

The data transfer is effected by means of signal transmitters and receivers. Apart from inductive steering (see later chapter 3) what is involved here is position information as well as commands which initiate reactions both on board the truck and at the floor installation. The number and type of com-

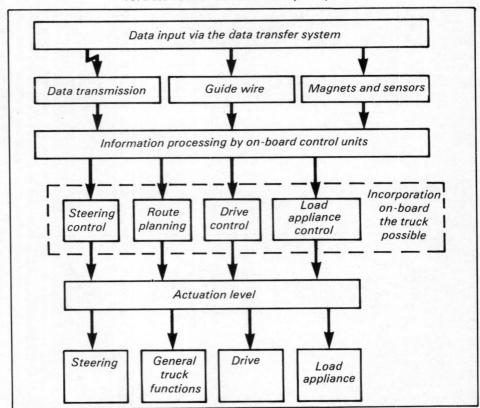

45: Data flow for driverless transport systems

COMPONENTS OF AN AGVS

ponents can be adapted to the operating conditions in modular fashion. Both truck and floor installation have active and passive transmission components at their disposal.

The principal ways of transferring information between truck, floor installation and the entity in charge (man or computer) are summarised in Fig. 46 and are described as follows.

For automatic operation the truck transmits switching commands to the floor installation. This is done by inserting 24V electromagnets which control reed contacts laid in the floor.

A distinction should be made between

☐ Section magnets for switching block sections, switching gate and traffic light controls and for switching stationary components (for example, chain conveyors for load transfer)

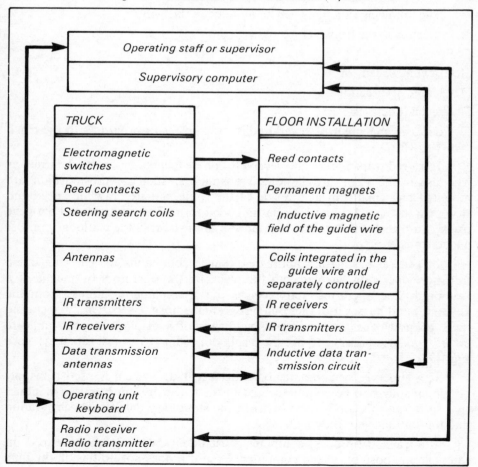

Fig. 46: Classification of the data transfer system

and

- [ ] Points magnets which use single frequency techniques method to switch the vehicle from the main course. In contrast to the section magnets which are in constant operation the points magnets are only activated for a short time. Special switches can also be controlled with points magnets (46, page 1304).

Permanent magnets are components embedded in the floor which control reed contacts located underneath the truck. Functions such as slow motion and stop are initiated in this way; in addition position information can be transferred for route planning.

Search coils for tracking the approximately 10kHz magnetic field used for inductive steering are located at the front of the truck and at the rear if a reversing facility is provided. Coils can also be found in the floor for transferring data to corresponding antennas on the truck. These coils are either incorporated in the guide-wire circuit or they are installed independently and are therefore separately controlled.

The following functions can be initiated in this way:

- [ ] Indicator on the truck, left or right
- [ ] Stop
- [ ] Crawling speed
- [ ] Change of direction
- [ ] Special functions
- [ ] Coded arrangement of several coils for transfer of programs to the truck (46, p 1507).

Infra-red transmitters and receivers can be located on board the truck as well as in the floor installation. Transmitters, such as those which are mounted on a platform at the side of the route, initiate crawling speed and stop or transfer a destination input in coded form. Transmitters on board the truck can transfer the destination address of the load to the stationary unit at a load transfer point.

In general destination addresses, that is commands for carrying out a transport operation, are fed into the truck by the operators by means of a keyboard. The operating unit receives the alphanumeric address from the keyboard and passes it to the route program memory. User-friendly operating units acknowledge the command by means of a display and can indicate errors. It is also possible to use the display to call up truck test functions – see Fig. 47.

If a supervisory computer or a truck calling system is used transport commands have to be communicated to the truck. In order to do this inductive data transmission is used between the stationary data processing station and the stationary or moving truck.

If a transport process computer is used then the truck will have to communicate its position to the computer. Even if it does not do this all the time

COMPONENTS OF AN AGVS

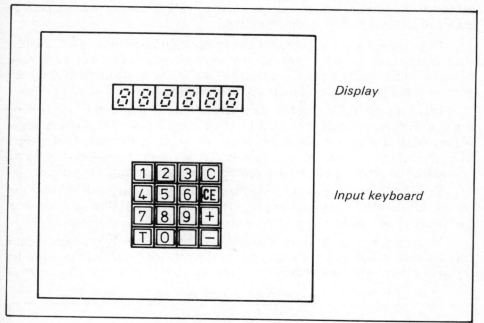

Fig. 47: Operating unit and display

then it should be done at least at certain points; it should also report on completed transport operations. This type of information transfer ensures that the best use is made in complex systems of the existing truck capacities by reducing the number of empty journeys and waiting time.

For duplex operation the transmission rates are of the order of 2,400 bauds (= 2,400 bits/sec) (5) and so are adequate for performing the following data transfer functions:

☐ Transmission from central control to the truck through the floor installation, (For example, go to destination 51, pick up the pallet on the right-hand side, go to destination 61, deliver the pallet on the left-hand side and return to waiting station 1) The extent of the data telegram determines for a given transmission rate the length of the loop laid alongside the guide wire. It can have a maximum length of 10m and is normally incorporated in the groove of the guide wire. The loop return is at a maximum distance from the guide wire of 1m (17, p 1529). The truck receives the telegrams by means of ferrite antennas (first frequency using simplex mode) and in duplex mode (second frequency) and is then in the position to report back to central control.

☐ Transmission from the truck to central control through the floor installation in the form of actual values, feedback information and acknowledgements. (For example, truck 6 with pallet 4711 at position 17. An error recognising code is preset for security of transmission).

It is reported (51) that data transmission is also possible through the guide wire of the inductive steering system without explaining the performance of this evidently very simple system.

Data transmission using radio allows constant communication between the truck and the control centre (usually a process computer). The advantage of using radio apart, from the constant flow of data, stems from the cost savings of not installing the data transmission loops.

However the main problem with radio transmission is the selection of suitable frequency bands approved by the German Post Office which allow reliable and interference-free radio communication in the industrial range; thus a satisfactory measure of reliability for this data transmission must be guaranteed by several transmission and receiver stations within the route.

In addition the central computer has to know the current position of the trucks. Either it takes this information from the floor installation control or each truck must describe its position with respect to certain landmarks (37).

In the latter case the truck makes use of the previously mentioned data transfer system between the truck and the floor installation. Data can also be transferred by infra-red transmitters and receivers although only over a

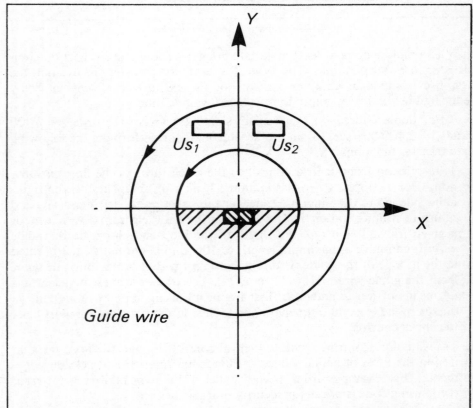

Fig. 48: Magnetic field around a guide wire and its scanning

transmitter-receiver distance of 10–15m (50). Despite adequate transmission security the short radio or infra-red transmission path hinders communication between the truck and a central computer.

The inductive steering principle was first described in detail by Schick (54). A further discussion of this principle can be found in Gunsser (29) who wrote that the guide path is represented by one or several wire loops laid in the floor. They are laid in grooves 6–10mm wide and 15–20mm deep which are cut into the floor and then sealed with synthetic resin. Fig. 48 shows the concentric magnetic field propagation around this guide wire when it is supplied with low-frequency AC.

On trucks with fifth-wheel steering the search coils are situated horizontally next to each other in the steering head. These search coils are induced by the magnetic field generated by the guide wire. The signal voltages U_{S1} and U_{S2} are separately rectified after amplification and led to a difference amplifier (comparator). If the truck is running exactly along the track the difference voltage is zero.

If the truck deviates from the track a difference voltage is generated as a function of the deviation which, as an amplified signal, produces a control signal in the steering motor as a regulating unit to initiate a course correction. The course which is now followed is maintained until more difference voltages are generated; that is, course deviations which make the next correction necessary. This completes the control loop of a constantly self-correcting travel direction which takes place inductively without any contact.

The steering can be effected also by two coils arranged vertically above one another and whereby comparison of the phase position of the two voltages decides the course correction, (29).

The reactions within the control loop happen so quickly that even narrow corners can be negotiated without any effort. Jerky steering can be avoided by soft and adjustable damping. Fig. 49 below shows the process schematically.

The guide wire is supplied from a quartz-stabilised low-frequency generator with power stage which with the help of the transmitter allows optimum matching to the length of the guide wire. The outputs are 15–20W, the voltages 20–50V and the currents 100–500mA. The frequencies used are 5.0 — 35kHz in accordance with the guidelines of The German Post Office. Fig. 50 below shows a block diagram of a frequency generator.

If different frequencies are required inside an installation the frequency spacing must be adequately dimensioned. In addition loop filters should be provided for common sections with different frequency.

For every additional frequency (4)
$$f_{n+1} \quad 1.2 \cdot f_n$$
where f_n is the frequency with index n
and f_{n+1} is the frequency with index n + 1 (= each additional frequency)

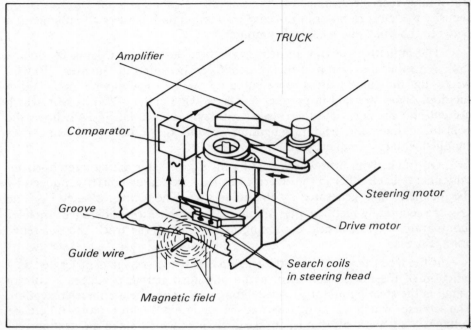

Fig. 49: Inductive steering principle

With simple round trips or circuits the current is controlled by a central transformer unit. In this case a frequency generator can supply up to 1,500m.

If the installation is divided up into block sections the current is regulated by the switching cabinets associated with each block. Due to extra conduction loses caused by section control the travel section which can be supplied is restricted to approximately 750m. Larger floor installations therefore require more frequency generators (46, p1506).

According to the VDI recommendation 3562 AGVS trucks are electric motor driven trucks with battery supply; DC series wound or compound wound motors are mainly used here. Electronic pulse control provides con-

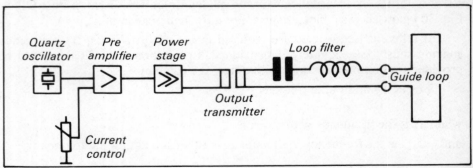

Fig. 50: Block diagram of the low-frequency generator for supplying the guide wire

COMPONENTS OF AN AGVS

tinuous and smooth driving and has won general acceptance since it also facilitates the positioning of the truck at load transfer points.

In the case of emergency braking with some trucks in addition to electromagnetic braking the drive motors can be switched to generator operation which produces a further braking effect. Fig. 51 below shows the block diagram for a drive control system with speed control and maximum current limitation.

The load transfer control depends on how load transfer is effected physically. If load transfer is to be automated it will require an appropriate control system.

Insofar as only the truck is active during load transfer (such as, forklift trucks, skid tractors) the load transfer can be incorporated in the driving program or the command is issued separately to the truck from a disposition computer over an induction loop.

If both truck and load transfer station are active during the transfer (such as through chain conveyors) the operation of the load transfer station

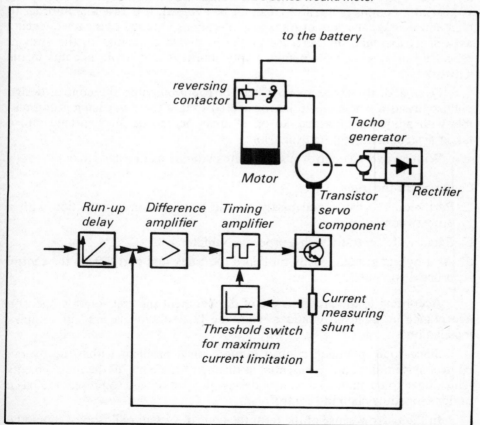

Fig. 51: Pulse control for a DC series wound motor

can either be initiated from the truck as a follow-on activity or centrally through the installation control system.

If only the load transfer station is active during transfer the same two options are open of control from the truck or from the installation control system.

The choice of control system for load transfer will always be linked to the choice of the overall master control system.

Route planning is of key importance within the framework of control activities. Route planning should be understood as the control system for empty and payload journeys or a combination of empty and payload journeys.

There are fundamentally different ways of arranging route planning control within the overall control design philosophy. These are essentially determined by the transport operations of the user on the physical and information levels and by the possibilities of the control system which the AGVS manufacturer can offer.

With complex transport operations, and in particular with production-integrated AGVS applications, the assimilation of the AGVS route planning system to a computer hierarchy cannot be avoided. It is essential in order to plan and manage the sum of transport operations in terms of their sequencing as well as to make the best use of the available capacities of the system, especially in seeking to minimise empty journeys and avoid pile-ups in the network.

Over and above the narrow field of route planning the control design philosophy as a whole has to be considered so that the sequencing is compatible with adjoining interfaces to other areas or can be incorporated into a larger system in any future expansion.

With an overall control system route planning can be carried out

☐ On board the truck
☐ Partly on the truck and partly on the track but in conjunction with a process computer
☐ Exclusively centrally with a process computer
☐ As a hybrid system, partly on board the truck and partly from the central process computer.

According to the present state of development the first solution has won acceptance in small installations while the last solution is used in complex installations.

Since route planning control is the central problem within the overall control system design it is essential at this point to define all the requirements which have to be made of a control design philosophy and to present the basic options for route planning control.

In the narrow sense of the term the control system philosophy should be

COMPONENTS OF AN AGVS

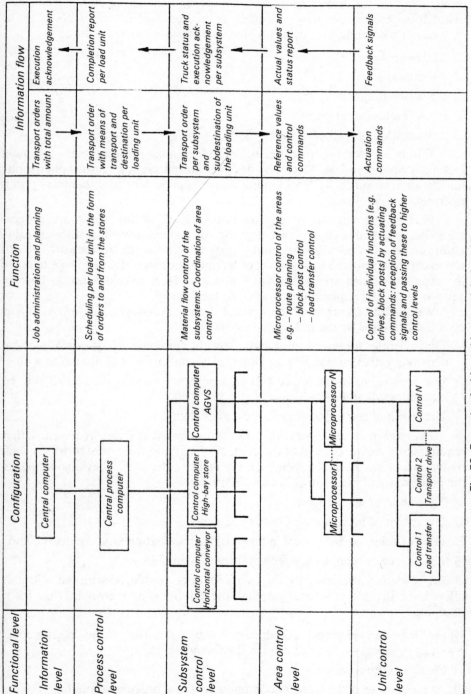

Fig. 52: Example of a hierarchic control design philosophy

understood as the implementation and the co-ordination of the truck and installation control systems. Insofar as higher-level activities have to be carried out by a computer these involve as transport process control

- ☐ job planning
- ☐ job management
- ☐ truck planning
- ☐ truck management and
- ☐ optimisation of transport operations

Within expensive computer-based information systems the computer hierarchy shown in Fig. 52 can be designed to represent a control philosophy for the application of AGVS in manufacturing processes or in complex goods distribution operations.

In this systems approach the lowest level of the transport process computer is linked to the actual truck control system, involving amongst other control systems the route planning system. Within the framework of this representation only the lowest level of a computer hierarchy need be examined in conjunction with an AGVS since all higher-level solutions are to be considered as special applications specific to the factory.

Within this framework the control design philosophy for an AGVS is essentially defined by the following factors:

- ☐ The existing transport problem with the parameters of material flow density, system complexity and internal organisation in conjunction with
- ☐ The constraints of the space and environment in which the AGVS is to be incorporated, and
- ☐ The required economic viability and availability.

These factors can be defined as the minimum that are required. Other requirements can be desired of the control system but the extent to which such requirements are fulfilled depends on the ideas of the user. They also influence the philosophy of the control system. They are:

- ☐ The extent of automation
- ☐ The extent of integration of data processing
- ☐ The flexibility of the system in terms of reorganisation and expansion, and
- ☐ The degree of redundancy with respect to availability.

The ideas formulated by the user should be compared with the solutions offered by the manufacturer so that the relationship shown in Fig. 53 is obtained.

Organising route planning. As already mentioned earlier in the chapter route planning control can be organised in the following ways:

- ☐ On board the truck
- ☐ At a point along the track in conjunction with a process computer
- ☐ Exclusively centrally in conjunction with a process computer

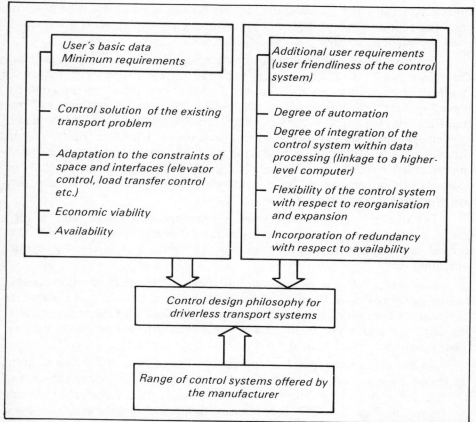

Fig. 53: Factors influencing the control design philosophy of driverless transport systems

☐ As a hybrid system, partly on board the truck and partly centrally in conjunction with a process computer.

The different solutions provided by these design options will be described below.

Route planning on board the truck. The on-board route planning system can be found in most of the systems which have been so far installed. The destination is either input through the operating unit on the truck itself or remotely whereby the orders are transmitted to the truck by call buttons at individual points of demand by a supervisor (head of truck operations) or also by a higher-level computer by means of induction loops.

With most of the installations which use a small number of trucks the expensive link-up of the individual truck control systems to a central computer can be dispensed with.

The advantage of on-board route planning lies in the control autonomy of each truck. This means that if the route planning control system on one truck fails all other trucks remain operational. Operator input of the destination is economic since the central computer is relieved and there is no need for communication. Thus on-board route planning control systems are used in most applications involving relatively few trucks.

The idea of an onboard control system can be extended by organising the trucks, not at the stopping point but centrally by a supervisor, that is by the head of trucks operations, who has a control console at his disposal. By means of data transmission loops the supervisor is in the position to communicate with the truck at certain points. In a few cases it may be sufficient if the central supervisor exclusively inputs the orders to the route control system through the operating unit on board the truck at a central station. The orders are signalled optically or acoustically by call keys in such a way that the individual points of demand have their own control systems and if necessary are processed in a counter unit.

The use of a central supervisor is recommended where there is high traffic density or where transport operations have grown so complex that central organisation is more efficient than exclusive local routing by the operating staff.

However, central supervisors are used also when the production process has to be controlled at the same time as the material flow so that a sequence of transport operations is produced which is identical with the order processing sequence and which in its turn is predetermined by production control. The systems are called central job distribution systems, (55).

As an aid to the supervisor for the specific control of the overall transport operation a route display panel which always shows the current state of the network can be used. The entire transport network is represented to scale as shown in Fig. 54. The individual sections are distributed on mosaic modules and LEDs indicate the position of trucks and in which direction they

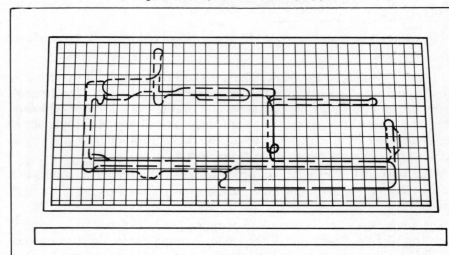

Fig. 54: Example of a route display panel

are going. The minute-by-minute display of the current situation as it is gives the supervisor the opportunity of intervening in unforeseen breakdowns (for example, pile-ups or prolonged stops). The route display panel presupposes an information transfer system between the trucks and the installation, that is it is necessary to make connections to all section control systems.

The supervisor is dispensable when the orders given at individual demand centres issued by call buttons are collected in a call memory. The call memory then passes the destination commands to the first encountered truck at a point on route. In this way priorities for the individual demand centres can be taken into consideration. This system is found particularly in simple allocation operations, for example in the assignment of empty trucks in a queue to individual commissioning areas.

A third way of extending on-board route control is to keep the control on board the truck but to carry out the transport process control centrally by means of a computer as shown in Fig. 52. The transport process computer then carries out the functions of the supervisor but it is also in a position to perform optimising strategies which are usually beyond the capabilities of a supervisor once the system has grown to a certain size.

With the help of process control it is also possible to take over the function of block post control centrally insofar as the communication between computer, route and truck is adequately dimensioned in order to control right-of-way among the trucks at any point.

Furthermore, there is the option in this example of a direct link-up with higher-level production control computers if on the process control level there are other subsystems besides the AGVS to be controlled in terms of material flow, see Fig. 52.

The advantage of this solution – on-board route control with central transport process control – lies in the hierarchical structure of the control design. This makes it possible for the system to be autonomous at every level and therefore remain available should the higher-level system fail. The separation of the route planning system from the transport control system brings the further advantage of separate introduction and testing which can considerably shorten the duration of the commissioning stage.

When designing the control structure care should be taken to ensure that each control function is as close as possible to the process itself; that is it is carried out on a lower level so that the amount of information transfer between hierarchical levels can be kept as small as possible.

Fig. 55 shows the options for on-board root control which also can be used in combination.

The location of the route control system on board the truck as illustrated in Fig. 55 determines at the same time the possibilities of addressing the destination after taking over the reference values (27).

In general the reference values are stored on information carriers. These are:

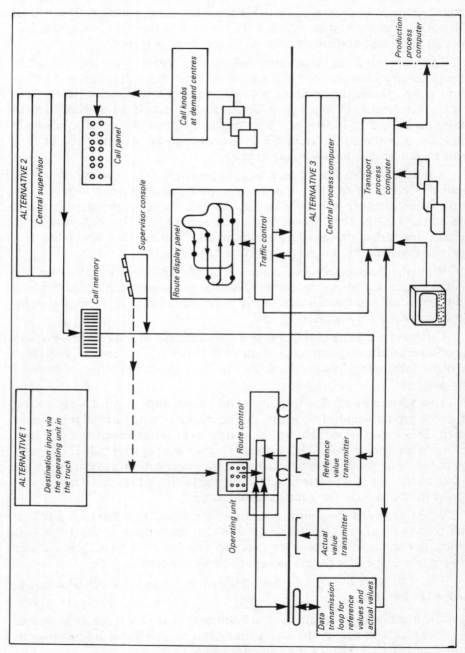

Fig. 55: Options for on-board route control

COMPONENTS OF AN AGVS

- Documents
- Documents which can be read by machine, such as punched cards
- Computer memories
- Electronic control systems

The reference values are taken from the information carriers into the control circuit so that the truck can drive to a destination. In the case of decentralised root control the function of taking over of the reference values remains on the vehicle. In addition the transmission of reference values becomes necessary when the addressing is carried out at other places than directly on board the truck, for example, by call memory, by supervisor or by computer.

In exceptional cases the reference values can also be received at the load itself (transport container). This can be worthwhile when the load is transported by a different means along its journey. Also, in this case the reference value has to be transferred to the truck control system. Insofar as alternatives 2 or 3 of Fig. 55 are implemented (remote reference value transfer) there is only reference value reception by the operating unit; this allows emergency operation by manual input of the destination or for mixed operating mode for example from goods-in to the high-bay store by manual input or from the high-bay store to commissioning areas by means of inputting the destination by inductive data transmission (transfer of reference values).

This example illustrates a further advantage of decentralised route control, namely the possibility of extending automation in stages by means of process control while keeping an operation emeregency mode through manual input. However for emergency operations to be carried out there must be information present in this state which permits input of the destination.

Decentralised route control at several places. Decentralised route control at several places attempts to make a compromise between the advantages and disadvantages of decentralised route control on board the truck on the one hand and of central route control on the other hand.

Decentralised route control on board the truck has the advantage that each truck is autonomous. However in large transport systems the costs multiply for route control (in terms of the hardware) as a function of the number of trucks. In comparison, the costs of a central control system when using a great number of trucks rise only moderately because only a central computer is involved.

This computer however must exchange information with all trucks at several places which with large networks produces considerable extra expenditure on the installations. This extra expense is kept as low as possible by having route control units stationed at various points in the network.

This system of stationary decentralised route control has been used, for example, in extensive, multi-storey hospitals. It needs usually a central process computer to manage the transport orders and monitor them,

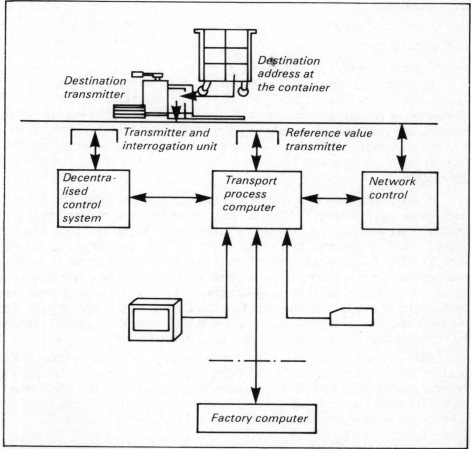

Fig. 56: Example of a decentralised stationary control system with central transport process computer

especially when lifts, horizontal conveyors and AGVSs are used in turn.

The destination address of the order (in the form of a bar code attached to the container) then accompanies the transport container and the trucks receive commands from the decentralised control systems at key points. The stationary control systems in their turn communicate with the central computer as shown in Fig. 56.

Insofar as the destination is inserted through punched cards on the transport container at individual stopping points this reference value must be passed to the truck transmitter and from there through the installation to the transport process computer so that the order can be activated from there.

Central route planning using a central process computer. With central route control a single transport process computer controls simultaneously, or nearly simultaneously, the routes of all trucks. At the same time the computer handles the transport operations of all the trucks in the network as well as the interfaces (load transfer, elevator sections, and so on). Steering control and drive control remain on board the truck itself. Load transfer, insofar as the truck itself is active during the operation is usually also controlled centrally from process control. The truck therefore just has small control programs at its disposal which are activated by the central computer.

This control concept assumes the computer is informed of the current state of the transport process at a large number of points in the network. At the very least this is necessary at points where decisions have to be made.

An important difference between the decentralised route control and the central process control (alternative 3 in Fig. 55) is that the central process control needs to retrieve actual values of information much more frequently since the destination control is not delegated.

On the other hand process control is well informed of the situation in the network at any point in time so that network control in the form of block control can be dispensed with because the computer controls the transport operations in such a way that given network sections can only be occupied by one truck at a time.

The Functioning of route control. All route control systems are based essentially on the reading of actual values which are compared with reference values. This comparison forms the basis for decisions which guide the truck step by step to the destination until the actual value agrees with the reference value.

If the route control system is on board the truck then in the case of microprocessor control the network is mapped in a semi-conductor memory. In general PROM memory modules are used, namely Programmable Read Only Memories which are used mainly in the EPROM version, that is as Erasable PROMs which can be removed using UV light (12, p. 175). Using this technique the memories can be reprogrammed and so can be adapted to future changes in the network.

All section features, such as stopping points and branches, receive code numbers which are issued only once over the whole network and so can be unambiguously identified (46, page 1310). However, it is also possible in principle that incrementally-coded sections are read and processed, as described in the following chapter.

After the destination has been inserted into the on-board computer reads the information from floor magnets located on the section. These are picked up through the truck-based receiver which compares this location value with the address or reference value and the EPROM content. In this way it can find out whether the truck can continue to take the main route or whether certain measures have to be activated, such as turning the vehicle into a branch route.

If the floor coding value and address value of the destination agree then the on-board computer immediately signals for the truck to stop. The diagram shown in Fig. 57 shows the route control principle as a data flow diagram.

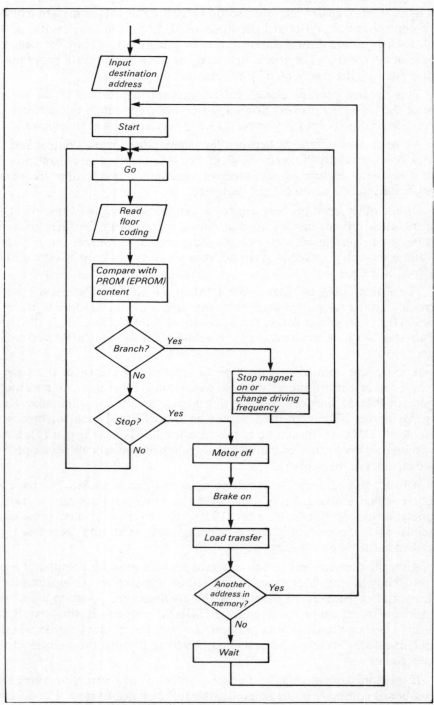

Fig. 57: Flow diagram for route control

COMPONENTS OF AN AGVS

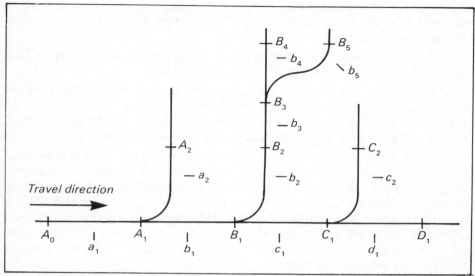

Fig. 58: Coding example of a travel route with incremental position recognition

Position recognition. Position recognition can be carried out in three ways independent of the number of frequencies used to supply the network as a whole – see later on in this chapter: incremental; digitally absolute; and digitally absolute and incremental.

With incremental position recognition the actual values are processed incrementally in the form of pulses. If the truck has driven past certain markers a_1, b_1 ... such as permanent magnets, (Fig. 58) a counter is started. The sum of the pulses corresponds to the distance covered.

If a current I with frequency f flows in the guide line passing through A_1–D_1 then the truck travels along this straight section until the counter has added up a number of pulses which corresponds to the reference value (address). If the truck is required to take the branch to b_2 then the control system has to give a signal to the stationary installation once the truck has passed the mark b_1; as a result of which the section B_1–C_1 is switched off and the current I with frequency f is switched to the branch section B_1–B_2 (27).

Instead of the principle of switching over points using a single frequency the multi-frequency principle can also be used for branching. The incremental system has a simple structure and is economical but in principle is more susceptible to interference and disturbances than the digitally absolute coded system. Networks with just incremental position recognition are only suitable for small systems since the capacities of the counting programs in highly branched networks rapidly becomes exhausted.

One of the problems is that it is very difficult to control a truck which, after having left a starting point and reached the first destination, has to go on to further destinations in another branch section; in which case it does not return to its starting station after each station for the counter store to be reset to zero. Also, after a temporary failure of the control system, or if the truck is removed from the network, the truck can no longer find its way using incremental position recognition.

For these reasons incremental position recognition if it is used at all is adopted now only in sub-areas of transport networks.

The second technique of position recognition is that called absolute digital position recognition. Here reference and actual positions, that is destination and key points, are represented as digital values in the binary system and compared in an arithmetic unit in the position control system.

This system is somewhat more expensive than the first system in terms of the number of magnets which have to be installed and in terms of the control design since not only do single pulses have to be gathered and processed but complete addresses too. However, it offers more or less total reliability of transport operations and has to be provided in complex systems as a matter of course. Even if a wrong decision has been made at one place this can be corrected at the next position. It is practically impossible for the truck to reach a wrong destination here.

In contrast to exclusive incremental position recognition programming can be considerably simplified if it is possible to group certain destinations into areas. In the following example, Fig. 59, these are denoted by areas A–D.

Using the area address the truck finds the branch area reliably and, using the destination address, the preselected destination. This is even the case for journeys from one destination to another which do not lie in the same area.

The two procedures described previously can be combined by separating out the individual transport areas to use absolute digital coding for principal routes and incremental position recognition to reach destinations along secon-

Fig. 59: Coding example for a network with absolute digital position recognition

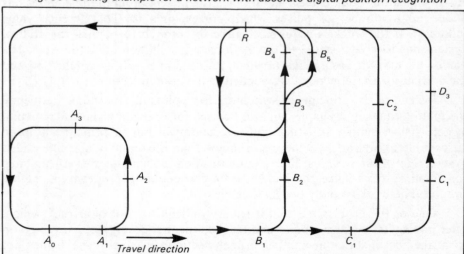

COMPONENTS OF AN AGVS

dary routes. In this way it is possible to combine the advantages of incremental position recognition (good value) with the advantages of absolute digital position recognition, namely, to limit programming effort to the necessary extent.

The network. The concept network is sometimes used synonymously with the concepts 'installation' or 'floor installation'.

Within the scope of this treatment the network should be understood as involving all stationary installations which in the strict sense are necessary for the inductively-steered truck to follow the given guide routes. According to this the network contains:

☐ The guide wire
☐ The floor-based and built-in installations for information transfer
☐ Traffic control.

To what extent interfaces such as load transfer equipment, elevators or automatic gates belong to the network cannot be answered one way or another. If an exact demarcation is required in terms of the respective suppliers then in the case of an automatic gate system the AGVS manufacturer, insofar as it is not acting as the general contractor, will supply the test section, the section switch and the gate control, including assembly; meanwhile the manufacturer of the gate system is responsible for the supply of the gate, the gate drive and the gate drive control system.

Turning first to the installation of the network the automatic and inductive steering of the AGVS truck makes it necessary to lay guide wires in the floor. The floor must meet the requirements specified in chapter 6. Particularly is this so when it is planned to use tractor trains, then it is recommended to trace the path with a prototype and mark the route for the guide wire so that the envelope curves are properly catered for.

A groove must be cut for laying the guide wire; this must have a maximum width of l0mm and depth of 20mm. The cutter, which is equipped with diamand cutting discs and is water cooled, requires a power supply of 380 volts and 25 amperes.

After cleaning and drying the joint the necessary guide and control cables are laid and the joint grouted with two-component plastic material, capable of taking loads at the latest one day later. All necessary coils, magnets, magnetic switches and other equipment relating to information transfer should be installed in the same way.

Since as a rule several trucks are being used, switching boxes for points, junctions and block posts have to be installed in the nearest wall or pillar. These control lines are laid depending on the local conditions on the wall or under the ceiling; wherever possible existing cable routes are used. Fig. 60 shows a cross-section through the guide wire whilst Fig. 61 gives an example of a complete network installation.

Networks can be supplied either by one or several frequencies within the

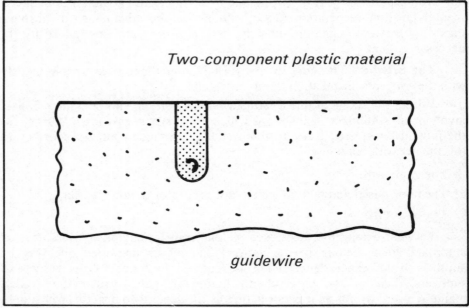
Fig. 60: Cross-section through a guide wire

guide routes. If several frequencies are supplied network areas are generated which can be used by the truck during navigation or branching. With the single frequency principle the redirection from the main lane is carried out by switching off the current in the main lane of the points block section. This is the basis for distinguishing between the single and multi-frequency principles.

In the past the technical arguments as to the advantages and disadvantages of the single and multi-frequency principles often played a key role in deciding for or against a certain manufacturer. From today's point of view this question is only of secondary importance because

☐ Both principles have basically the same cost with respect to the overall system costs
☐ Both principles can be essentially used in all networks
☐ The production of a software-based control program by converting the hard-wired logic to microprocessor control reduces the differences between the single frequency principle and the multi-frequency principle.

The next sections give more details on the differences between these two principles.

Network control. Network controls are designed as microcomputer systems in keeping with current developments and have the following basic functions:

☐ block control
☐ data transfer of actual values for the purposes of transport control

COMPONENTS OF AN AGVS

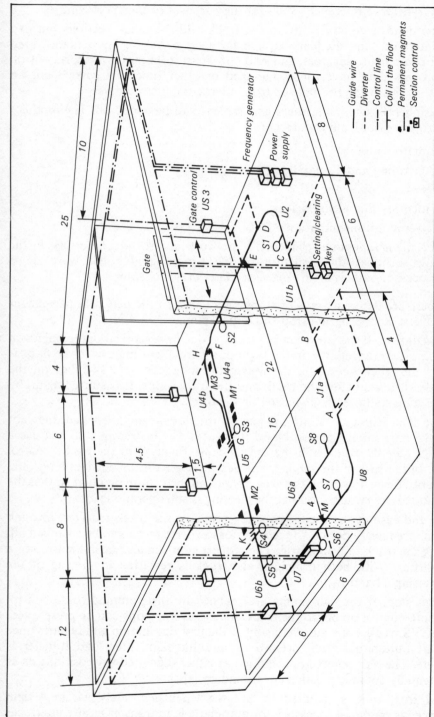

Fig. 61.: Example of a complete network installation using the single frequency principle

☐ data transfer of reference values for the purposes of transport control

The control not only covers the truck and the block sections but can cover interfaces and auxiliary equipment such as traffic lights, gates, load transfer equipment, elevators, and so on. If, apart from pure transport control, store process data have to be managed or other functions carried out an integrated system can be made up from microcomputer modules.

Peripherals such as printers and screens can be added on. The modules for microprocessor traffic control are:

☐ The microcomputer
☐ The operating and display unit
☐ The power supply
☐ The memory module as well as
☐ Receiver and transmitter units

The main functional elements of network control are described in the following sections; the structure of the programmable hardware on a microprocessor basis is briefly discussed later in this chapter.

There are different ways of stopping a truck on the network: permanent stop, request stop, program stop and optical stop.

Permanent stop. If the truck always stops at one or several stopping points on each run a permanent stop in the form of a permanent magnet in the floor is used. If the truck goes past the maget the magnetic field switches on the magnetic switches below the truck and the truck stops. It is started again by the operating staff pressing a button on the truck.

Request stop. The request stop is part of the floor installation and does not need any extra equipment on board the truck. The operating staff can use a switch to halt the truck utilising the blocking function of the traffic control system. In the case of the single frequency principle this only leaves a residual current of 20mA flowing in the area of the switched stopping point so that the drive motors are switched off but the steering is still operational.

In the case of the multi-frequency principle the truck receives, through the block frequency device, the request for the drive motors to be switched off, bringing the truck to a standstill. The request stop can also be in the form of a light barrier. With both principles the truck is restarted by freeing of the section using a hand switch.

Program stop. If the halt points are reached by programmed route control with route control on board the truck one or several stops can be preselected. The AGVS truck stops automatically at the first destination and restarts once the start button has been pressed or automatically after load transfer is complete. The halt points are identified as either digital coded positions or as incrementally counted positions in secondary lanes.

Optical stop. If it is required to halt the truck accurately within a tight tolerance, as might be necessary for automatic load transfer at any interfaces

COMPONENTS OF AN AGVS

or when joining lifts, then the precise position can be achieved using photoelectric cells positioned on the truck. This is especially so if the total weight of the truck changes. The required positional accuracy can only be achieved if the speed is reduced in good time and floor conditions remain unchanged.

If several trucks are used within a network then clearly the traffic has to be controlled. Traffic control can be achieved by a decentralised route control system based on the block switching principle or centrally by a process computer.

If a process computer is used to control large complex systems then all trucks can be interrogated in rotation to determine their position and, depending on the traffic situation, receive stop or go commands. Process computer control requires adequate position coding in the network as well as data transfer to the process computer using inductive devices or radio control.

A description of conventional switching techniques using decentralised route control is outlined as follows but first a distinction has to be made between two blocking methods: buffering blocking within straight sections, and crossing and junction blocking. (The term blocking is similar to that used in railway signalling). With crossing and junction blocking only the fixed block sections are defined by means of switching points. Block control is carried out by the route control system using interactive data transfer between the truck and the network installation. The basic framework of the block circuit is the same for the single frequency principle as it is with the multi-frequency principle; however the method of operation is different, as discussed later.

Buffering blocking is designed to prevent trucks from colliding with one another as they form a queue on straight sections. As a rule fixed block sections are defined as they are with crossing and junction blocking. With straight unbranched sections however it is also possible to provide interlocks for the trucks and this is done entirely from the trucks themselves. In this case an adequate distance is maintained between trucks, for example, by infra-red transmitters and receivers.

This method has proved itself for straight, unbranched sections. The advantages are:

☐ The blocking distance between trucks can be reduced
☐ The maximum transport performance for this section is increased
☐ Trucks can be buffered within a confined space at depot stations.

The blocking gap between trucks can be set to within controlled limits by suitable adjustments at the truck.

A similar buffering system can be adapted using ultrasonics in which case the truck behind sends out an ultrasonic beam to the truck in front. Compared with infra-red blocking there is a number of additional advantages:

☐ The broad ultrasonic beam can even recognise trucks in front on bends
☐ The load on the truck in front can be used as a suitable reflector

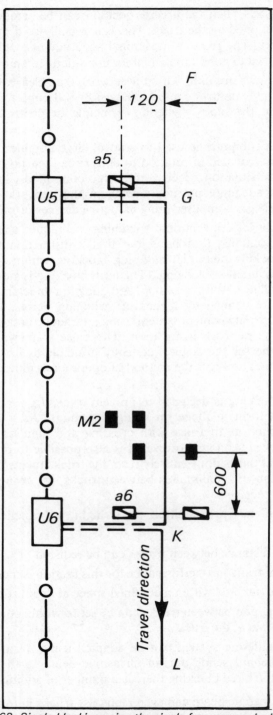

Fig. 62: Single blocking using the single frequency principle

COMPONENTS OF AN AGVS

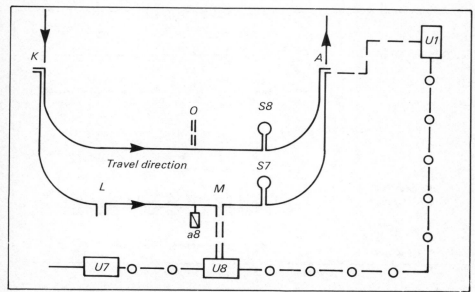

Fig. 63: Blocking of a confluence using the single frequency principle

Buffering blocking is not necessary when very low speeds are encountered (V < 0.5 m/s). In this case the trucks are not maintained at a distance but each truck coming from behind makes contact with the truck in front and switches off its drive by the impact with its bumper. In this case the trucks are designed so that the rear truck hits the truck in front at a suitable place.

If fixed block sections are used then blocking control using the single frequency principle and the multi-frequency principle are carried out in different ways.

In the case of blocking using the single frequency principle when the installation is switched on all block sections are supplied with AC power of 10kHz and 20mA. This allows the steering to remain operational although the trucks cannot proceed because of the electronic interlock circuit. By inserting a block-setting key the current in one section is increased to 100mA allowing the first truck to be started. When the truck moves into the next block section the magnet situated beneath the truck switches a reed contact in the floor requesting release of the block section.

As shown in Fig. 62 the truck energises the magnet switch a_5 which in turn switches on the section G–K in the control system U_5, blocking the section F–G lying beyond. In the dynamic case a truck can only enter the section G–K when the truck behind has left the section K–L and so has blocked the section K–L but has released the section G–K.

At crossings or junctions a two-way blocking is carried out which, as shown in Fig. 63, has the following form. The truck's section magnet activates the magnet switch a_8 on moving from L to M and switches on the section M–A through the control system U_8; at the same time it blocks the section

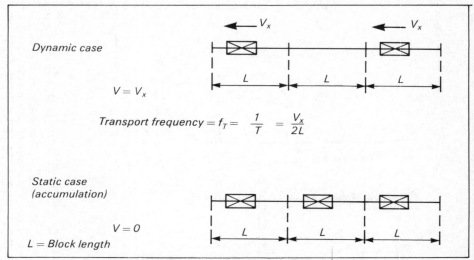

Fig. 64: Dynamic and static cases for block control

L–M. In this case K–A is also blocked to prevent collisions at point A. S_7 serves as a signal for switching on the truck's direction indicator. With high traffic densities K–A will be divided in the middle at O; then K–O forms a double block together with K–L in the form of a turnout whereas O–A forms a double block with M–A in the form of a confluence junction. In this case trucks can already enter the section K–O when a truck is in the section M–A at the same time.

On straight sections, because of the effectiveness of the block control system, the dynamic case can be distinguished from the static case. In the dynamic case the moving trucks are kept at a distance of one free block from one another by the block control system; in the static case each block section is occupied by one truck as the result of queue formation, as shown in Fig. 64.

In the case of blocking using the multi-frequency principle the basic principle of alternate releasing and blocking of block sections used in the single frequency principle is retained. However with the multi-frequency principle all block sections are continuously supplied with current. The blocking system therefore does not act on the interlock circuit by switching off the current but uses its own blocking frequency to block and release the sections.

The induction loops of the block sections are supplied with approximately 30kHz and blocking control proceeds as shown in Fig. 65. The truck 1 in front drives past the currentless induction loop 2; because it does not receive the blocking frequency it can proceed. On leaving the area D–E the induction loop 2 is switched on and the induction loop 1 switched off by passing the switch b; so the sub-area D–E of the entire block D–F is blocked while the entire block B–D is released.

The control logic can also be formed in reverse, so that driving is possible

COMPONENTS OF AN AGVS

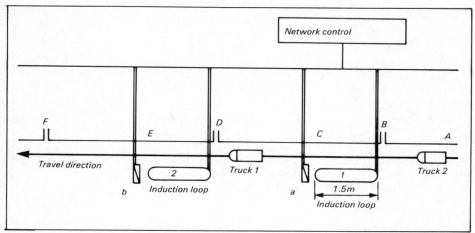

Fig. 65: Block control using the multi-frequency principle

with current in the block loops. In this case, with a breakdown of current supply (fault), drive is automatically cancelled and an additional safety measure is considered.

The block control can also be based on a system of receiving and transmitting loops using the multi-frequency principle whereby no high-level network control is necessary, for example as shown in Fig. 66.

In this case each truck continuously sends out a signal to the receiving loop which then passes it to the transmitter. Using the blocking frequency the transmitter then blocks the transmission loop behind. This signal is immediately used as a command to stop by the truck following behind.

The flexibility of this special block control system is that at any point trucks can be removed from the network and replaced. This is particularly useful when during commissioning trucks running an errand with the driver on board can return to their destination station without driver but using the guide wire.

In the same way as with the single frequency principle parallel junction

Fig. 66: Flexible block control using the multi-frequency principle

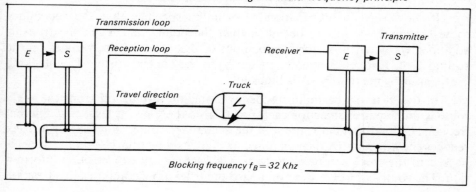

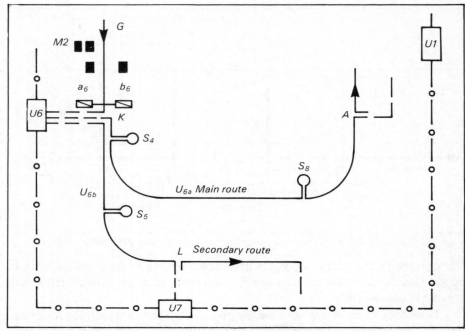

Fig. 67: Points switching using the single frequency principle

sections are blocked off at junctions and crossings. In this case priority can be given to the first truck to arrive, or the main routes can be given priority over secondary routes by means of a special signal.

Points. The use of points to turn from a main route to a secondary route is fundamentally different for the single frequency and multi-frequency techniques.

Using Fig. 67 it can be seen that for single frequency operation when the magnet coding M_2 is passed a check is made as to whether the main route can continue to lead the truck to the preselected destination. If this is the case the truck's section magnet activates the magnet switch a_6 in the floor; the control system U_6 switches on the section K–A and blocks the section G–K using block control.

If, however, the destination programme specifies that the secondary route section K–L has to be taken then the points magnet in the truck is additionally switched on for a maximum of 3 seconds. This switches on the section K–L through a magnet switch b_6 in the control system U_6, blocking G–K; at the same time K–A is blocked.

In the case of the multi-frequency principle turning off from the main route is much easier since for each points area there are at least two different frequencies for the main route and the secondary route, (Fig. 68). If a truck reaches the key point A then an absolute digital signal (mark 'a' in the floor) is passed to the truck's control system as an actual value of position information. The route control system on board the truck now compares this position

COMPONENTS OF AN AGVS

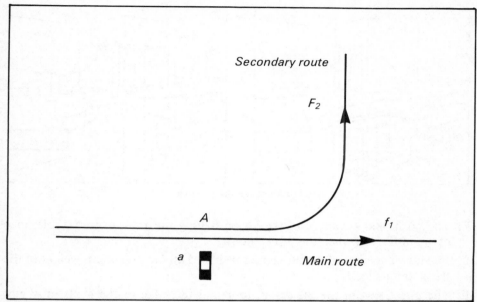

Fig. 68: Turning off from main routes using the multi-frequency principle controlled from the truck

information with the stored network and the destination to be reached (reference value). If the destination is to be reached through the main route the frequency 1 remains set in the selective amplifier of the steering control. If it is necessary to turn off from the main route then the selective amplifier of the steering control on board the truck is switched over to frequency 2.

Discounting the position information, this means that branching off from main routes using the multi-frequency principle is completely controlled from the truck.

Gate control. If a truck moves towards a gate which is automatically controlled, the truck's section magnet first passes over a floor magnet which, through a unit in the truck's control system, provides the signal to open the gate. While the gate is opening a section immediately in front of the gate is blocked. When the gate is completely open a limit switch is actuated causing the section immediately in front of the gate to be released. Once the truck has passed through the gate section the gate drive mechanism is activated to shut the gate with the help of a magnet switch in the floor. If there is enough room at the location the switch for opening the gate acts sufficiently early to allow the truck to pass through the gate section without stopping.

Lift journeys. Lifts can be run with either forward and reverse travel or merely with forward travel, (Fig. 69). In either case the following requirements are made of lifts:

☐ The lift must be barred to other traffic

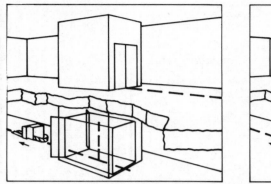 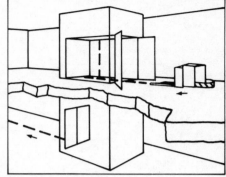

Fig. 69: Lift journeys with through travel

- The positional accuracy of the lift on halting must not exceed a tolerance of ± 5mm, even when the truck enters.
- The load capacity of the lift should be based on the maximum weight of the truck and its load
- The gap between the lift cabin floor and the edge of the shaft must not exceed 10mm
- A recess has to be provided in the lift to accommodate the guide wire
- A trailing cable has to be provided for the guide wire connection between the shaft and the lift cage.

During automatic journeys the truck calls the lift and waits for it in an interlock section in front of the lift. This section is released when, after the lift doors have opened, the lift is ready to go. The truck moves into the lift at a very slow speed and is positioned exactly by photocells. This monitoring equipment gives the signal to the lift to start.

The interlock circuit remains active throughout the journey. At the same time the truck informs the lift control of the destination floor. When the doors open at the preselected floor the command to leave the lift is given to the truck through a limit switch; switching units in the floor inform lift control that the truck has left so that the lift is now ready to undertake another transport operation.

In order to provide a fully automatic transport system automatic load transfer stations are essential. The basic requirement for these interfaces is the compatibility of the load units of the truck and the station as well as that of the control signal exchanges which are usually carried out by the network control system. The exchange of signals between the truck, the network and the load transfer station is usually carried out inductively or with the help of photo-cells. The control sequence is essentially the same as at lift and gate interfaces.

Truck calling. Trucks can be called using one of a number of techniques, including inductive data transmission, coils or luminous displays.

COMPONENTS OF AN AGVS

With inductive data transmission the truck is called by depressing a call key, usually sited at the point of demand. As soon as this call has been registered by the central computer a signal lamp in the keyboard is illuminated. The light goes out as soon as this transport command has been issued to a truck at a data transmission loop.

With coils a calling key is also activated by the illumination of a signal lamp. At a defined point in the network a combination of coils is switched on which, as a code, corresponds to the calling address. Should a truck which is not in use pass this arrangement of coils then it automatically takes this information as its new destination address. Immediately afterwards the call is acknowledged by means of a magnet in the truck and a magnet switch in the floor. The display light goes out, reporting the arrival of a truck.

When luminous displays are used to call trucks there are in principle no interfaces with the network. The call is made by pressing a call key associated with signal lamps on the request panel. The operator or supervisor at the panel dispatches a truck to the request station and the call is acknowledged by pressing a key. This simple technique also can be implemented by the operator himself (46, page 1531).

Programmable hardware. Microprocessors make it possible to process instructions and to carry out arithmetic and logical operations. Fig. 70 shows a block diagram for a microprocessor. The memories belonging to the microprocessor are working memories; the microprocessor does not have an

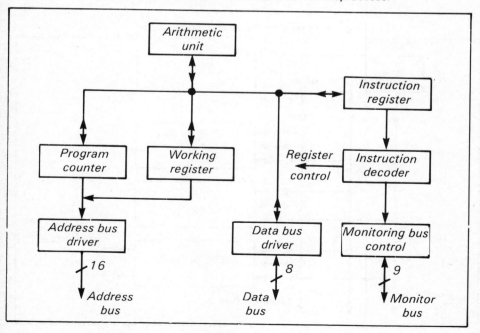

Fig. 70: Internal structure of the MC 6800 microprocessor

input/output unit at its disposal. Consequently microprocessors cannot work by themselves (12, p526).

In order to be able to operate a microprocessor requires an external memory containing the sequence of commands to be performed, such as the program. This memory can be a RAM (Random Access Memory) in which case the commands are read after switching on. If the program does not have to be changed it can be stored in a ROM (Read Only Memory). Insofar as occasional alterations need to be made to ROMs they are replaced by the so-called EPROM (Erasable Programmable Read Only Memory).

Special interface circuits are used for the input and output of data. In this way signals from switches, electro-magnets and contactors are transferred to and from the arithmetic unit by a PIA (Peripheral Interface Adapter) while data associated with the operating unit or inductive data telegrams are transferred by means of an ACIA (Asynchronous Communications Interface Adapter). PIA modules enable the parallel (simultaneous) transfer of data between computer and outside equipment (peripherals) whereas ACIA modules transfer the data serially (successively).

A complete system made up in this way is called a microcomputer. It is shown schematically in Fig. 71; the microprocessor here is the central control and computing unit (Central Processing Unit, CPU).

The block diagram (Fig. 71) still does not say anything about the performance of the CPU and the capacity of the working memory. It is simply a block diagram of a computer. Computers can be classified into a number of sizes:

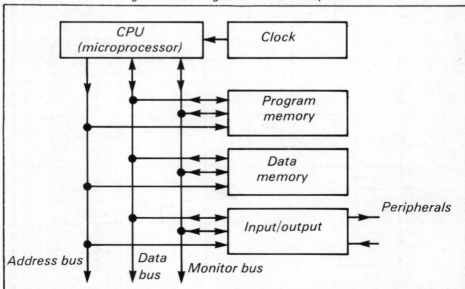

Fig. 71: Block diagram of a microcomputer

Main-frame computers: over 265K words at 24–64 bits
Minicomputers: 8–256K words at 12–16 bits
Microcomputers: 0.5K–64K words at 4–16 bits

Microcomputers made their breakthrough with the advent of monolithic microprocessors. With their rapidly falling prices their applications in equipment development became of interest where they were usually in the position to undertake relatively complex computing and control functions using fixed programs. In this way a standard hardware circuit is produced for many applications, whereas the actual development work (specific applications) is transferred to the production of the programs or software (12, p527).

Microcomputers are successful mainly because far more functions with a much greater complexity can be offered at a favourable overall price compared with previously hard-wired logic.

The advantages of programmable controllers compared with hard-wired logic are shown in the following table. (30, p89).

PLANNING	– higher complexity possible
	– standardisation of hardware
	– shorter time for producing control functions (software)
	– complete documentation
	– standardised interfaces to higher-level process computers
	– more reliable feasibility studies on the control system – better offer price
MANUFACTURE	– pre-manufacture or extensive deliveries of manufactured components
	– short and improved tests at the manufacturing factory
	– high quality
COMMISSIONING OPERATING and SERVICING	– higher reliability
	– easy modifications on expanding the range of the application
	– simplified fault finding
	– usually shorter repair time and re-commissioning

Table 1: Advantages of microcomputer control compared to hard-wired logic control

However the positive effects of the use of microprocessors is not only to be seen in truck control systems but in the field of high-level transport process control.

Increases in the performance/price ratio for microcomputers have meant that since about 1978 programmable controllers have replaced hard-wired (fixed programmed) control systems as truck control systems. Conventional control systems are now only sometimes offered for very simple networks with a few trucks (22).

With the help of the microcomputer it is now possible to transfer a large number of the high-level control functions to the lower process level. This has the advantage that the expenditure involved in the information transfer and frequency of information transfer can be cut down. The sub-systems, thanks to a relatively high intelligence, can be adequately designed to operate under emergency conditions in the event of the central transport process computer failing.

Within a scheme of a step-by-step expansion, a complex system can be started without a central process computer and it is also possible to structure the entire computer hierarchy in such a way that the bottom control level already consists of its own independently functioning modules.

After being incorporated in the truck control system microprocessor technology has also found its way into network control.

Chapter Four

Applications for the AGVS

SINCE their introduction in 1955 the areas of application of AGVSs have widened significantly. AGVSs are now represented in all branches of industry and trade. Restrictions result mainly from the payload or the tractive power of the trucks and the dimensions of the goods to be transported. Restrictions can stem also from the existing spatial conditions or the transport needs which, in quite specific applications, demand an overhead transport system.

For these reasons AGVSs are not used at the moment in process manufacturing industries and heavy industry.

Insofar as the application of an AGVS is technically feasible its actual implementation in competition with other transport systems is decided from economic considerations based on investment calculations as well as from technical considerations which together with the results of an economic analysis can be summarised in the form of an efficiency analysis (10).

With complex systems or a large investment it is advisable to separate the efficiency from the cost evaluation and to use as the basis for the decision a cost-effectiveness analysis, (41).

Requirements for the use of AGVSs will now be described together with examples of applications; the economic aspects will be dealt with in chapter 7.

In (25) nine criteria were named in a check list favouring the use of an AGVS in comparison with other transport systems. If at least five of these criteria are satisfied then it seems worthwhile to look more closely at the possible use of an AGVS in a specific application.

The idea behind the concept of check lists and the cost comparison curves shown in chapter 7 is that the considerable planning effort, which is often involved, should be made only if there are well-founded reasons for the feasibility study.

The criteria of the check list are as follows:

- More than 5% of the daily throughput which is handled by fork-lift trucks or other driver-operated vehicles is delivered either to the wrong place or is late in arriving at its destination.
- There are at least 10 load transfer points in the transport system under consideration.
- The total throughput between the load transfer station is $>$ 35 load units an hour and $<$ 200 load units an hour.
- The savings potential in the transport system is equivalent to at least three forklift trucks in each of two shifts.
- In the present factory layout there is at least 100m of installed roller conveyors or power and free conveyors
- The production control system in the factory under consideration relies on the on line processing of information on materials movements
- Materials, which are either deposited in gangways or next to machines, hinder access to individual work places and lower productivity because of the need for additional handling.
- The material flow is to be linked automatically to a high-bay store or to a high-level process control system
- The transport goods are sensitive to impact and need careful and precise handling.

Distances and material flow. Out of the large number of AGVSs which have been installed to the present day, it has been found in practice that the minimum length of the network is 100m; very large networks have a total length of 3,000 to 4,000m.

The key variables in deciding the capacity of a transport system based on an AGVS are:

- Truck speed
- Block lengths
- Load transfer times
- Switching times
- Availability of the trucks or of the overall system
- Scheduling of journeys
- Topology of the network

Since the orders and therefore the truck movements in the network are subject to stochastic influences only a simulation of the overall transport process can give information on whether transport capacity under certain conditions can be achieved.

Using the example of a simple circular network with a source of material flow (input) and a delivery point for the material flow (output) the theoretical maximum capacity is obtained for transport of load units using fork-lift trucks when the minimum block distance is 10m and the truck speed is 1m/s.

In this case, (Fig. 64) in order to achieve maximum capacity each block is occupied by a moving truck whereby at the destination a truck emerges from the last block with a load unit once every 10 seconds, that is 360 trucks or load units an hour.

This theoretical maximum transport capacity is reduced by waiting and switching times on the network, as well as extra time for load transfer. The result is that a maximum of 250 load units per hour can be achieved in a network, given that there are enough trucks available. With tractor operations, because of the more favourable utilisation of the network, a maximum transport capacity of approximately 400 load units (pallets) an hour can be achieved using four trailers equipped with a total of eight load units.

Possible applications. The use of AGVSs can be divided into four main areas of application:

☐ Supply and disposal at store and production areas
☐ Production-integrated application of AGVS trucks as assembly platforms
☐ Retrieval especially in wholesale trade
☐ Supply and disposal in special areas, such as hospitals and movement of files in offices.

The first area includes many different applications in the entire internal materials flow from goods-in, through the store areas and production sections right up to dispatch. Here the AGVS takes over the functions of collection and distribution or both functions at the same time.

With regard to the high level of organisational integration in the overall system of operations the following special applications have become particularly important:

☐ Servicing high-bay stores with central process control.
☐ Servicing of engine test cells in the car industry, especially in the electrical test station as well as in performance testing.
☐ Servicing of production work places according to the principle of central work distribution.
☐ Integration of the AGVS in packaging and palletising processes, such as travel through shrink-wrapping ovens and automatic palletisers.
☐ Integration of the AGVS in the servicing of flexible manufacturing systems.
☐ Use in areas having a health hazard, such as cold stores or areas with the risk of radiation.

The second application area for AGVS trucks used as assembly or manufacturing platforms is based on the idea that the manufacturing process can be structured in a more flexible way with this transport system than is possible with conventional continuous conveyors.

The assembly process, which takes place on the truck as the workpiece carrier, can then be carried out fully automatically. Typical examples include welding lines or assembly operations where there is no need for synchronisation or where work places can be controlled individually by the truck. In the latter case typical examples are the assembly of car engines and gearboxes, or the final assembly of electrical equipment to car engines. Through the individual work of the assembly workers it is possible to combine the production units into larger units and so carry out more tasks than had it been conventional assembly line.

The third application area can be found in order-picking for wholesale trade. Here there are three potential uses for AGVSs. In the high-performance area, where there is a rapid turnover of articles, retrievals are taken only from the bottom level with the retriever moving next to the truck because of the high access rate. The truck is controlled by the retriever.

In the second performance range, which involves articles with a medium turnover rate, the access rate is low therefore larger distances have to be covered from one article position to another. For this reason the retriever travels on the truck. In addition, depending on the product range and the amount of available space, the truck is equipped with a lift platform in order to reach the second rack level, (Fig. 29).

The third possibility which occurs with product ranges with medium and low turnovers is to use truck tractors with roller containers, (Fig. 23). In this case the order-picker starts by filling the last container and in so doing stands between the last and last-but-one roller contained on the truck. After retrieving the last container he climbs over the last-but-one container and pushes it to the back. Now the order-picker can service the order for the last-but-one container. Since with this narrow aisle retrieval system it is impossible to exit from the side of the truck in the case of an emergency the way ahead is always kept clear so the truck can be crossed with the aid of a step. The retrieval of the other five or six containers takes place in the same way.

Supply and disposal of materials. In an enterprise which manufactures electrical equipment the AGVS can be used for the transport of material between goods-inward (area 4) and the store (area 1) on the one hand, and between the store and the production regions (areas 2, 3, 5) on the other hand (Fig. 72). The disposal of material from the production section is carried out by another transport system.

On a network having a total of 700m length, seven sets of points and 21 stopping places there are four electric driver-seated tractors with hand pallet trucks as trailers. In single-shift operation approximately 400 pallets are handled daily using this equipment. The pallets contain small and medium-size electrical parts in containers.

The AGVS does not have any station at its disposal in the sense of a central point from which the tractor trains can be organised. Instead of this the trucks are addressed with the destinations of the current orders by means of the on-board keyboard and they are started from one of the stopping points in areas 1–5.

APPLICATIONS FOR THE AGVS

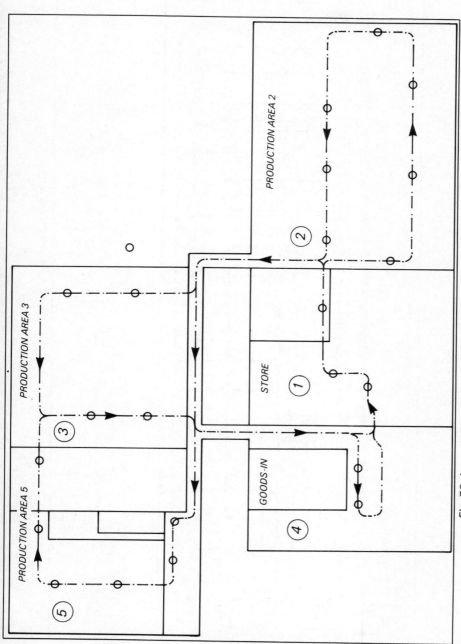

Fig. 72: Layout of the network in an enterprise of the electrical equipment industry

AUTOMATED GUIDED VEHICLES

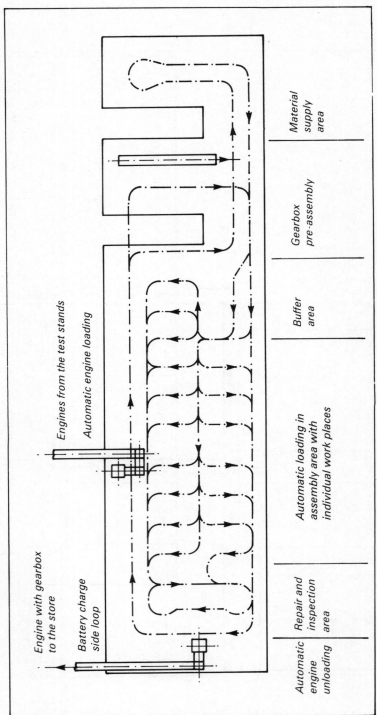

Fig. 73: Layout of the network for the engine assembly application

So that every operator addresses the truck correctly the keyboard panel to input destination information carries a plan of the destination points. The truck stops at the destination address and waits until it is loaded or unloaded by hand. When the tractor train has been serviced a new destination is fed in and the truck is started. Since only small containers of the same type were used automated load transfer was dispensed with. If the pallets are densely loaded then in some cases the pallet trucks are uncoupled from the tractor train, driven manually and then coupled back again. Apart from the 21 fixed stopping points the truck can be stopped at any point on the network by manual operation of the stop switch on board the truck.

The floor installation is based on the multi-frequency system and equipped with the normal blocking system. The network is designed as a pure linear path control system; it therefore controls the priority of trucks at junctions and crossings as well as the distance between trucks. The total of five areas can be reached by three different frequencies.

Production-integrated applications. During final assembly an AGVS is used as an assembly platform to supply work stations in a car factory. Assembly of the engine and gearbox is carried out at 16 work stations. The network, with a total length of 500m, uses 35 trucks to assemble 800 engines a day in two-shift operation.

After the engines have been loaded using a vertical conveyor the trucks have a choice of one of two loops in which they are supplied with the corresponding gearboxes (Fig. 73). Loading is achieved without pallets, but with the gearboxes 'attached' to the engine with two bolts. Other small components are found at the individual assembly stations.

After attaching the gearboxes the trucks enter a buffer area. This area can take a maximum of four trucks. The trucks are then called to one of the 16 assembly stations. The trucks, Fig. 74, are equipped with three drives: a travelling mechanism, a steering drive and a hydraulic drive. The hydraulic drive is put into operation during assembly so the operator easily can gain access to all sides of the product.

Power is supplied from a battery (24 volts, 75Ah) with a capacity designed for approximately two shifts.

After completion of the assembly operation the trucks move on to the load transfer station. This area has a secondary loop to permit the changing of batteries. The trucks drive automatically into this side when battery state is low. There is also another loop for repair and inspection work on AGVS trucks.

Control of the entire assembly system is split into three levels. The top level is a scheduling computer acting as a freely programmable control system. The second level involves the network of the AGVS manufacturer's control system. The route control is decentralised and carried out on board individual trucks.

At the load transfer stations photo-electric cells position the trucks with

Fig. 74: Final assembly of an engine with the help of AGVS trucks

an accuracy of ± 5mm. Within this positioning area the approach speed is 0.1 – 0.2m/s while on a free section the trucks can travel at a maximum speed of 0.5m/s.

In the search for the best transport system the choice fell on the AGVS although it was relatively the most expensive in terms of the investment costs. On the other hand a reduction in assembly costs was achieved against the previous synchronised assembly line operation. The time savings for each assembled engine stem from the fact that with different types of gearboxes the conveyor had to be timed to accept the engine with the most time-consuming gearbox. Using the principle of individual work stations these buffering times are eliminated and so up to five minutes per engine can be saved. A smaller part of the time saved comes from improved accessibility of the engine on the AGVS truck.

Interfacing with high-bay stores. The automated and economic link-up of production areas on both sides of a high-bay store serviced by stacker truck and stacker cranes is especially important for large factories in order to

APPLICATIONS FOR THE AGVS

achieve the degree of automation in horizontal transport which has already been obtained in high-bay store technology. At the same time this fills the gap which has prevented a computer-based systems approach to material flow control. The interface between the high-bay store and the horizontal system is the forebay of the store whose system elements are defined in the VDI recommendation 2690 (Fig. 75).

The transport system for in-feed and out-feed of load units between the high-bay store and the production area must meet the requirements of both sides. According to (31) the following requirements can be made of the transport system as a link-up unit:

☐ Matching of the transport performance to the maximum performances of the storage and production systems.
☐ Adequate buffering facilities in order to uncouple the system elements.
☐ Positioning within specific tolerances at the load transfer points.

The type of store, the throughput, the transport paths, the required degree of automation and the nature of the transported goods are the main factors which determine how the store forecourt is organised for material flow. There are various ways of organising the transport operations and five different types are discussed according to (31).

Type 1: Only conventional industrial trucks operate in the forecourt, such as fork-lift trucks and pallet trucks. The pallets are set down on stands at floor level. The number of collection points for in and out-storage is the same or smaller than the number of store aisles. The pallets are taken into the storage area by the stacker crane from a stand on the floor or directly from the floor by a stacker truck by means of swivel reach forks.

Type 2: Fork-lift trucks in the forecourt area take the pallets from a driverless transport system or another automated transport system. After identifying the load the truck driver is directed to the store aisle and the pallet is there set down on the floor, stand or conveyor system.

Type 3: The collection points for in and out-movement in the storage area are accessed directly by the automated transport system and the load units set down at floor level. The number of collection points does not have to be the same as the number of aisles. In the simplest case there is one collection point per aisle. No additional transport equipment is necessary for buffering; stacker trucks transfer the load units to the storage area with the help of swivel reach forks.

Type 4: Each aisle in the storage area has its own transport equipment. This equipment serves either as an in-storage or out-storage buffer, bringing the load unit into the right position for the stacker crane or the stacker truck which can be equipped with a telescopic fork for handling the load.

Type 5: In front of the racking aisles of the store there is an in-storage system with a loading area, profile control, I-point and the connecting conveyors which carry out the distribution to the individual aisles. The out-storage

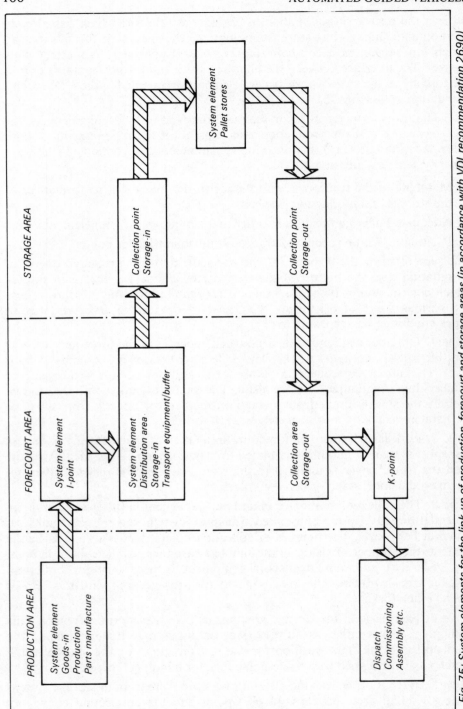

Fig. 75: System elements for the link-up of production, forecourt and storage areas (in accordance with VDI recommendation 2690)

APPLICATIONS FOR THE AGVS

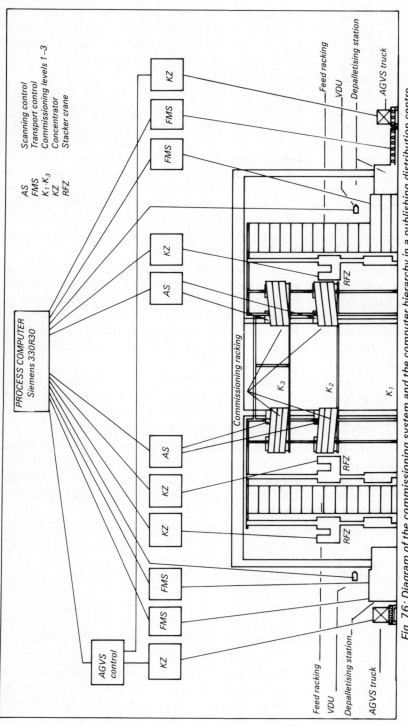

Fig. 76: Diagram of the commissioning system and the computer hierarchy in a publishing distribution centre

system consists of accumulation and path elements with a control point at the end. In the simplest case a sampling section is attached at the control point which terminates at the point of transfer to the horizontal transport system. With complex systems there are several transfer points present. A typical solution of this type of interface is described below.

In the distribution centre of a publishing business the AGVS links the high-bay store as a reserve area with the order picking area which has its own truck-operated feed system (Fig. 76). If the contents of one of the compartments in the racking has reached the refilling level the process computer initiates an outfeed of the corresponding article in the storage racking. At the outfeed conveyor in the high-bay store an AGVS truck automatically picks up the pallet and transports it to a depalletising station (Fig. 37) from which the corresponding number of articles is taken to a feed unit. After manual handling the discharged pallet is taken automatically again by the AGVS truck and transported to the I-point of the high-bay store where the pallet is reshelved as a new pallet into the high-bay store.

In the final stage of expansion the high-bay store has approximately 60,000 pallet positions and eight delivery conveyors to transfer loads to the AGVS. Transport between the high-bay store and the commissioning area is carried out by a total of 40 AGVS trucks which circulate on a network approximately 2,000m long and with about 100 stopping places and 50 sets of points. In single shift operation an average of 2,000 pallets can be moved in a working day.

Turning now to the organisation of the material flow the basic command which the process computer can give to the AGVS control system is as follows: transport order No. 4711, for pallet No. 4712, from position No. 1 to position No. 2, with priority 4. Other data telegrams are given in Fig. 77.

On receipt of such a command the AGVS truck takes the corresponding pallet from the load transfer station with the stationary conveyor. Both are active. The truck starts automatically with both the destination and the route control provided centrally. Only small routine operations are performed by a small microprocessor on board the truck.

Because of the high degree of centralisation of the overall material flow the trucks do not have their own keyboard for the inputting of commands. After delivering the load to the destination (depalletising station) the truck receives a new transport order. As a rule a discharged pallet from another depalletising station has to be brought back to the high-bay store. The transport order for the return journey is initiated at the VDU of the depalletising station by acknowledging that a certain number of articles have been removed from a pallet. Following this the process computer independently puts in a request for a truck for the return transport.

The AGVS control unit consists of a PDP 11/04 process computer with a VDU screen to display the network, the positions and the current status of the trucks. There is also a printer to record system errors. The conveyor

APPLICATIONS FOR THE AGVS

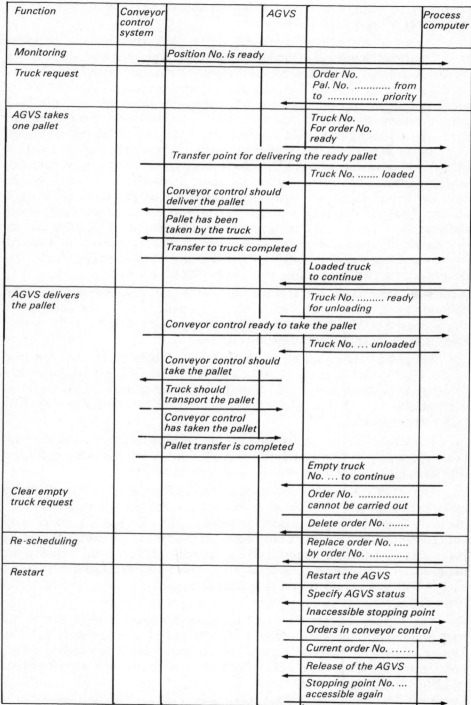

Fig. 77: Examples of the structure of data telegrams between conveyor control, the AGVS and the central process computer

control system is on the same control level and, as a contactor-relay system, is in charge of the load transfer stations.

The microprocessor on board the AGVS truck controls the drive operations, the load handling equipment, the roller conveyor and the steering motor. Ancillary functions, such as checking the charging state of the battery are also undertaken. A total of eight devices with microprocessor technology are installed underneath the floor which can communicate with the DTS control system, the conveyor control system and the trucks themselves. The data are inductively transmitted between the truck and the electronic device. For this an induction loop is installed in the floor with a transmission frequency in the vicinity of 100kHz at every key point which can be points or stopping places. The truck receives the most up-to-date information on the network through these induction loops by means of the respective devices and so can find the destination preset by the process computer after route optimisation without itself having to possess information on the route and the destination. By means of central route planning there is no need for a special traffic control system for blocking the trucks and it is also possible to bypass queues or to systematically free any queues which have formed.

AGVSs in warehouses. An AGVS is used in the central warehouse for a chain of food stores. This warehouse delivers on a daily basis to the outlets. The stores consist of 33 single racking aisles; each racking aisle has eight storage levels and 48 racking compartments (Fig. 78).

Altogether about 4,200 articles (including both food and non-food articles) are stored in some 12,500 pallet positions. Of these approximately 1,500 pallet positions are located in the bottom racking area and are therefore accessible for AGVS order picking. The other pallet positions are serviced by rack entry modules.

The total of 16 rack entry modules are responsible for all storage operations, the supply of feed pallets to the AGVS areas, and the order picking of articles in the top rack levels which have a low turnover.

The AGVS area consists of eight loops, (Fig. 78) and accesses every other aisle in the store. The aisles in between are serviced by a rack entry unit. Some 30 fork-lift trucks are used for order picking. Europallets, which have lattice frames and are used as the loading unit, offer the advantage over roller containers of having a larger carrying capacity. Although the vehicles are designed for trailer operations (that is, one hand-operated fork-lift truck for each AGVS truck) only the truck is used there are no trailers; therefore only one pallet is used.

The system runs on the multi-frequency system, has eight points and four programmed stopping places and a length of approximately 1,250m. During single shift operation, and a regular peak load in the early morning hours some 30,000 packages are order picked every day with a performance of 120–300 packages per man hour.

When an operator has completed a loop of the network corresponding to

APPLICATIONS FOR THE AGVS

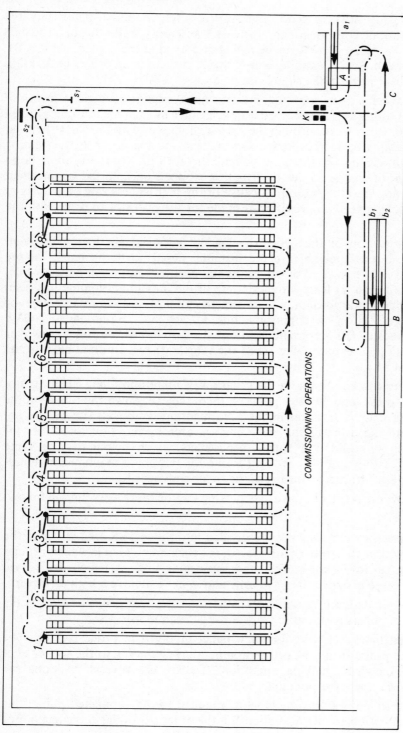

Fig. 78: Layout of the network in a commissioning system in the wholesale trade

an order picking area he sends the AGVS truck to the dispatch bay by pressing a button. Only then can he request an empty truck through the call button (Fig. 77 No. 1–8). The truck call is displayed at the control cabinet S2 and passed to the truck standing empty at S2. If there is no order the fork-lift trucks wait in a queue at the signal station S1.

In the order picking loop the operator always takes the articles from the bottom level of the two racks opposite each other in each aisle. If the articles are far apart he can work through the loop without stopping the truck; here he has the possibility of overtaking the moving truck on the left or on the right or to work immediately in front or behind the fork-lift truck. With a large number of positions per order the truck is stopped and started again when necessary using the stop button. Because of the high access rate of articles in this order picking area no provision is made for the operator to drive with the truck.

On completion of the order the loaded truck is sent to the dispatch department. Immediately before the dispatch department the height of the loaded pallet is interrogated by an automatic profile control. All pallets which satisfy the criterion 'at least three-quarters of the maximum volume has been reached' arrive at the automatic load transfer point D where a roller conveyor takes pallets to the foil wrapping machine for sealing. The unloaded AGVS truck continues to the automatic load transfer point A at the roller conveyor a_1 and picks up an empty pallet. Then the vehicle travels automatically to the order station S1.

All other trucks for which the criterion of minimum filling height is not satisfied go to the load transfer point C; there they set down their load on the track and take up a position in the lane mentioned above. The pallets are brought from the track to the consolidation area by hand-operated fork-lift trucks. After consolidation with other pallets which belong to the same dispatch order they are also brought to the foil wrapping machines by hand-operated fork-lift trucks by means of the roller conveyor b_2.

The flow of records in the order picking can be represented in the following manner:

- ☐ The orders for each dispatch are divided into the AGVS area in which the articles with high turnover are stored, the rack entry area in which articles with low turnover are stored, and the drinks area in which drinks are stored – these occupy a large volume and in general have a high turnover.
- ☐ A computer determines the likely volume of the split order as well as the number of pallets required to work through each loop.
- ☐ A printed label is issued for each pallet involved in order picking. The labels are printed with the article number corresponding to the sequence of steps in the order picking operation. If there are several positions per article there are correspondingly more labels.
- ☐ A red pallet number on the label is attached to the completed pallet; this makes it possible to check at point B if the order per outlet is complete. For

APPLICATIONS FOR THE AGVS

this the supervisor has a list showing the pallet numbers for each outlet.

Because of the simple network and the state of the organisation at the time of construction the route control system on board the truck is hard-wired using conventional technology. Only at two points in the network, S1 and K, are destination commands transmitted inductively to the truck. The call memory S2 transmits transport orders to the trucks, with certain high-speed areas having priority.

The conveyor control system is activated at the interfaces by means of photo-cells at the load transfer points.

Chapter Five

Experience with guided vehicle systems

EXPERIENCE with driverless transport systems has been assessed mainly by a market analysis carried out by (22) in 1979 in which 137 AGVS users in the Federal Republic of Germany participated by means of a detailed questionnaire. Of these 92 users responded to the questionnaire, corresponding to a 67% return. In addition 20 on-site visits were made.

It should be borne in mind that since the market analysis was completed major inadequacies have been removed mainly as the result of further developments in electronics.

Causes of failure. Since the number of users maintaining exact statistics of breakdowns is extremely small only a quantitive assessment could be made of availability. According to Fig. 79 it was found that 98.4% of the users judged availability to range from the very good to satisfactory.

An analysis of breakdowns for all the driverless transport systems in the study showed that 59% of all breakdowns resulted from truck electronics. One basic cause of failure of the electronics is temperature sensitivity. Due to the generation of heat by the electric motor and the battery in the truck a thermal load can be built up at summer temperatures which with compact trucks cannot be adequately dissipated (15). Additional fans installed in compact trucks can help here. Breakdowns due to mechanical failure were responsible for 15% of the reported cases.

Some 30% of users found failures due to the electronics in floor installations, 37% of users found failures due to the electrical system and 45% of the users found failures due to breaks in the conductor cable. In this connection it must be pointed out that approximately 50% of all systems and trucks are maintained by the manufacturer, while 68% of users have a utilization of 70% of their system. Some 30% of the systems operate on two-shifts while 6% operate on three-shifts.

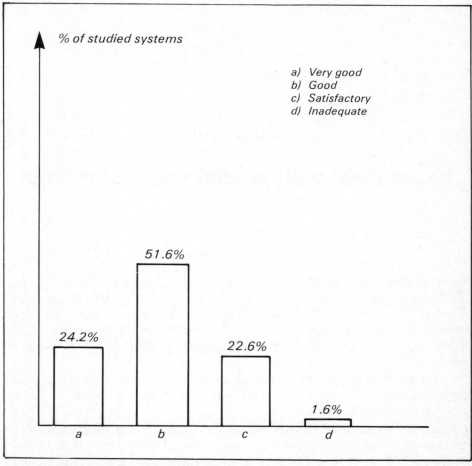

Fig. 79: Assessment of the viability of AGVS by the users of all the systems studied

When interpreting these results it should be borne in mind that the user's assessments do not correspond to the criteria which would allow an exact evaluation of the availability. The results should rather be taken as a starting point for making further improvements to guided vehicle systems.

Since making the market study a large number of short-comings has been removed, especially to the truck and network control systems.

An evaluation of the availability of electronic components taking the exact measurement specifications into consideration is presented in (5). The control of a technical system is seen here as a continuous process and the following relationships apply when for a large number of components Z_0 a number of intact components $Z_{(t)}$, and a number of failed components $F_{(t)}$ are observed at every point in time.

(4) $\qquad Z_0 = Z_{(t)} + F_{(t)}$

The reliability function and the failure probability as well as the failure rate can be derived from this

(5) $\quad p_{(t)} = \dfrac{Z_{(t)}}{Z_0}$ \hfill Reliability function

(6) $\quad q_{(t)} = \dfrac{F_{(t)}}{Z_0} = 1 - p_{(t)}$ \hfill Failure probability

(7) $\quad \lambda_{(t)} = \dfrac{dF_{(t)}}{dt} \dfrac{1}{Z_{(t)}}$ \hfill Failure rate

where Z_0 is the total number of observed components
$Z_{(t)}$ is the number of observed intact components
$F_{(t)}$ is the number of observed failed components
$p_{(t)}$ is the reliability function
$q_{(t)}$ is the failure probability
$\lambda_{(t)}$ is the failure rate

The availability of a system is defined as:

(8) $\quad \eta_s = \dfrac{MTBF}{MTBF+MTTR}$

where η_s is the availability of the system
MTBF is the Mean Time Between Failures and since it is the average time between two failures it is a measure of the trouble-free time.
MTTR is the Mean Time To Repair and therefore is the average time necessary to carry out a repair.

These two variables determine the availability such that with very long trouble-free times and very short repair times the availability asymptotically approaches the value 1.

Table 2 gives failure rates, MTBF, MTTR and availability for components, sub-systems and systems. It is found that with increasing complexity the MTTR increases and the MTBF drops because the failure probabilities of sub-systems within a system under consideration linked serially together multiply, thus the failure of one sub-system eventually causes the whole system to break down.

By constructing hierarchical systems with independent sub-systems and building in redundancies at critical points it is possible to improve the availability of the overall system and to obtain a relatively good level of availability, as shown in table 2.

Truck batteries. Apart from special cases traction batteries used in AGVS trucks weigh 200–770kg. Considering a cross-section of batteries used the power density with respect to 1kg of battery weight is on average 25Wh. This average value is based on PzS batteries which are designed for 1500 discharge cycles and can be discharged to a residual capacity of 20%.

The literature is full of newly developed batteries which have a considerably improved power density. Thus, for example, a 'cold' nickel-zinc

Components	(10^{-9}/h)	MTBF(h)	MTTR(h)	n	Examples
– Diodes					
– Transistors	5–10				
– Integrated circuits (SSI/SMI)	5–100				
– Resistors	100–200				
– Capacitors	1–10				
– Inductive resistors	10–1000				
– Transformers	10–30				
– Relays	10–50				
– Plugs	500–1000				
	10–300				
Assemblies					
– Digital input					
– Digital output	≃ 10 000	≃ 100 000			
– Memory card	≃ 7 000	≃ 150 000			
– (4k, 8 Bit, EPROM)	≃ 10 000	≃ 100 000			
– CPU (Intel 8080 including add-ons)	≃ 5 000	≃ 200 000			
Control equipment consisting of:					
– Complete chassis					S + S
– Supply unit including converter	180–200 .10^3	5000–5500	0,5–1,5	0,9997–0,9999	SESTEP 513
– CPU (8 Bit)					
– System card					
– Memory 4K RAM (8 Bit) 4K EPROM (8 Bit)					
– I/O 64 dig. Inputs 32 dig. Outputs					
Process computer consisting of:					
– Complete chassis	300–500 .10^3	2000–3000	1–2	0,9990–0,9997	Ferranti A700 E
– Supply unit + converter					
– CPU (16 Bit)					
– Memory 64K/16 Bit core memory		1000–1500	0,5–2	0,9980–0,9997	DEC PDP8
– Bus system	700–1000 .10^3				
– Asynchronous/serial interface					
– Teletype					
Mainframe computer consisting of:					
– Medium-size computer including 3 peripherals and standard software (USA study)	30 000–40 000 .10^3	250–300	2–3	0,9981–0,9934	IBM 360/30

Table 2: Failure rate, average troublefree time, average repair time and availability of electronic assemblies, sub-systems and systems

traction battery was presented in (28) which produces 70Wh/kg. In (16) a 'hot' sodium-sulphur battery is reported which produces 100Wh/kg at operating temperatures of 300–350°C.

Since however higher-performance battery types are still not sufficiently developed for mass production it can be assumed that designers of AGVS trucks will be faced for the foreseeable future with the conflict between compact and weight-saving construction on the one hand and the large battery capacity and working range of the truck on the other.

A certain improvement in the working range while reducing battery weight can be achieved when the organisation of the transport system allows

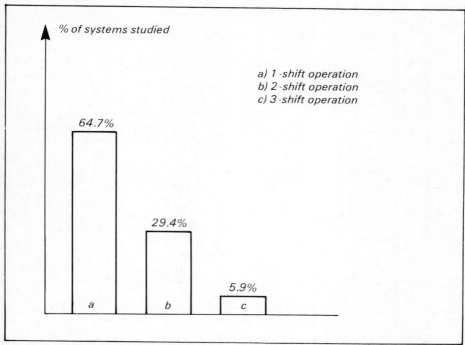

Fig. 80: Operating period (relative frequency) of the studied systems

Fig. 81: Battery charging procedure (relative frequency) in the systems studied

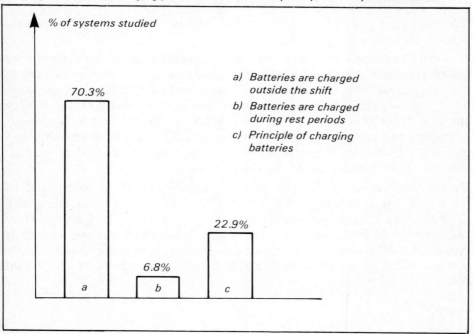

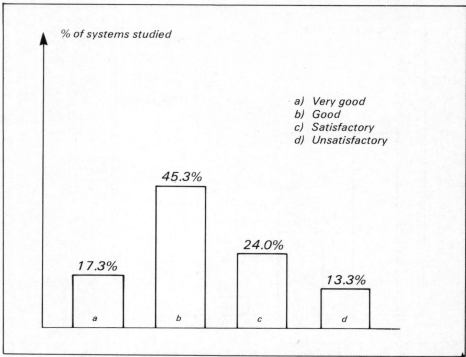

Fig. 82: Assessment of experience with AGVS by users of operational and shut-down systems

that the batteries can be re-charged by fixed current collectors at stations where trucks are waiting in queues.

User experience indicates that battery capacity is usually designed for the duration of a shift in order to prevent deep discharging causing damage to the batteries and to ensure trouble-free operation of the truck. If the operation period (single-shift and multi-shift operation) is compared with the battery charging procedure, (Figs. 80 and 81) then it is obvious that in single-shift operations the battery is charged outside the shift, whereas in two-shift and multi-shift operations the principle of changing batteries is used or the batteries are at least recharged during rest periods.

Overall assessment of AGVSs. Fig. 82 shows an overall assessment of AGVSs by users, including those which shut down their systems. According to this about 87% of all users judge their systems to be very good to satisfactory. 13.3% could not obtain any satisfactory results with their AGVS. A further analysis of this class shows that 70% of the unsatisfactory systems had been badly planned and 80% of these systems had been shut down.

Chapter Six

Planning for AGVS implementation

AS A RULE it is difficult to improve the material flow in existing organisations since in most cases there are relatively few opportunities to reorganise existing installations or to recover the costs involved.

Despite the low installation cost involved in introducing an AGVS to an already existing system it should not be forgotten that automated systems require more space than driver-operated industrial trucks in order to satisfy the safety requirements associated with accident prevention. Another set of problems involves the physical and organisational interfaces of the previous material flow. This means that for new installations to be successful the load transfer stations must be designed in such a way that either the staff operating them are fully extended or that load transfer is made fully automatic right from the start.

The survey of AGVS users (22) shows very clearly that the adaptation of the AGVS to the organisation as well as the problems of interfacing arose foremost in the planning, (Fig. 83). In this connection the survey of about 800 organisations is also significant. These organisations could count as potential users but had not yet installed an AGVS (22). It was found here that more than half of those questioned said that an AGVS could not be introduced within the existing organisation, (Fig. 84). These evaluations make it clear that when designing complex systems only with very thorough planning is it possible to achieve the full advantages of an AGVS, namely an automated and at the same time flexible organisation of the material flow.

Analysis of flow. The physical material flow consists of the following components:

- Type of transported goods (or load units)
- Order of transport operations
- Quantity framework of the material flow
- Distances of connections within the transport network

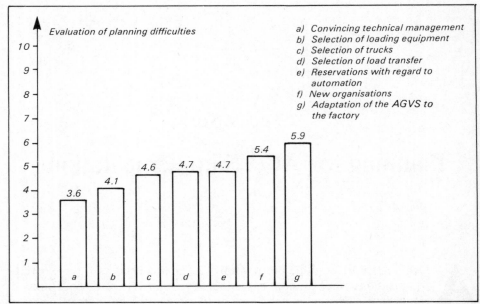

Fig. 83: Planning difficulties for AGVS from the user's viewpoint (arithmetic mean of the given evaluation grades 1–10; 1 = no problem, 10 = most difficult)

Fig. 84: Reasons for not introducing AGVS (relative frequency) in past projects

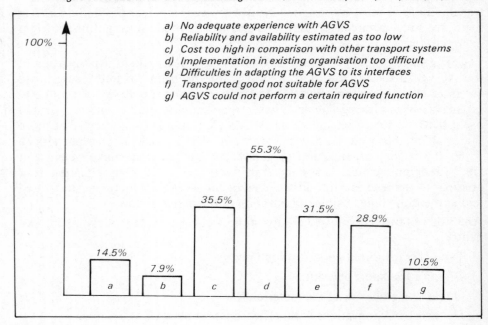

PLANNING FOR AGVS IMPLEMENTATION

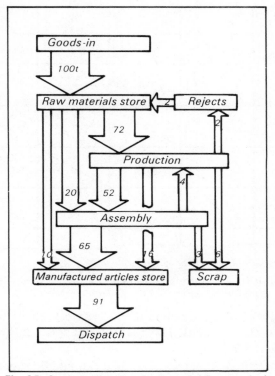

Fig. 85: Sankey diagram of a material flow analysis (Amounts in load units/time unit)

Within the framework of a new design consideration should be given to the standardisation of loading units insofar as this is still possible, as well as the opportunity to optimise the order of material flow. The quantity framework is best expressed in terms of a from/to matrix with allocation of the corresponding distances, (Fig. 85).

One form of representation in abstract form is the Sankey diagram, (Fig. 86). Alternatively the material flow connections can be drawn on to the original layout using arrows whose width corresponds to the transport quantity. Quite often the results of such a visual aid are sufficient to make certain improvements immediately obvious.

Space and floor conditions. The requirements made by an AGVS in terms of space and floor conditions depend on the type of truck, any gradients which have to be negotiated and on the type of floor installation used by the manufacturer. In each case the recommendations given by the manufacturer as minimum requirements have to be followed. As a matter of principle the following criteria must be satisfied:

☐ Transport lanes or gangways should be designed with a width such that there is a minimum clearance of 1m to the doors.

☐ It is recommended in tractor operations but also with single trucks to first

FROM \ TO		No. 1 Goods-in	2 HRL	3 Dept. A	4 Dispatch
No. 1	Goods-in		$\dfrac{X_1 \text{ LE}}{Y_1 \text{ m}}$	$\dfrac{X_2 \text{ LE}}{Y_2 \text{ m}}$	
2	HRL			$\dfrac{X_3 \text{ LE}}{Y_3 \text{ m}}$	$\dfrac{X_4 \text{ LE}}{Y_4 \text{ m}}$
3	Dept. A		$\dfrac{X_5 \text{ LE}}{Y_5 \text{ m}}$		$\dfrac{X_6 \text{ LE}}{Y_6 \text{ m}}$
4	Dispatch				

x: number of load units   y: distance in m

Fig. 86: Combined matrix for representing the quantity flows in load units (LE) per time unit and the associated transport distances (m)

follow through confined sections of the network so as to determine the space requirements of bends. This precaution applies in the first instance to fifth wheel steered trucks.

☐ The areas in which AGVS trucks circulate should be kept in a clean condition so that transport lanes remain dry and nonskidding. Metal rails, shafts, cover plates, recesses and other metal objects must not be situated in the transport area since they can interfere with inductive steering or inductive data transmission.

☐ Grooves should be cut in the finished floor covering. The edges of these grooves must not be allowed to deteriorate.

☐ The load bearing structure, the transition layer and the wear surface must provide an adequate electrical conduction resistance; metal pipes and reinforced concrete iron must if necessary be carefully and adequately earthed.

☐ Since inductively steered trucks are true-tracking high requirements are made of the scuff resistance of the wear surface; on the other hand a minimum static friction of $\mu \geqslant 0.6$ is required.

☐ The load bearing structure of the floor should be designed so that unavoidable joints, such as those caused by floor plates, cross the transport lanes at right angles; under no circumstances should they run parallel to the lanes. The expansion behaviour of the floor is the main cause of cable breaks. In certain cases compensation can be provided by double insulated cables or expansion loops in the cable.

☐ Unrelated electrical lines can interfere with the network control system and for safety reasons should be at least 0.5m away.

- [ ] The surface pressure of the floor covering must be equal to the truck's weight under all climatic conditions. The evenness of the floor covering is of special importance, in particular when fork-lift trucks are used. In these cases the AGVS manufacturers require that DIN 18202, page 3, line 3 is observed.

Network planning. There are no general rules for planning the network. With small networks the organisation of the transport lanes does not present any difficulties apart from the limitations of space. However, the situation is quite different with complex networks where peak loads occurring at certain times and/or in certain places can easily lead to queues and bottlenecks in capacity.

Since every network represents a special applications problem in itself only general guidelines can be given for planning networks and avoiding bottlenecks. In exceptional cases it may be necessary to simulate the limiting capacity as well as local bottlenecks.

After determining local bottlenecks an attempt can be made to increase the capacity at the bottleneck using the following indications. Using an iterative procedure it is possible to improve the capacity of the network in stages.

- [ ] Networks consist of a set of closed loops in which all loops have a connection with the main loop. Reversing is only possible in exceptional cases, for example in spurs, otherwise the entire network is governed by the one-way traffic ruling.
- [ ] Secondary lanes carrying a high material flow should not join the main lane which is also running at high density in the same area. If it is impossible to avoid a confluence in regions of intense material flows the block distance should be reduced to a minimum. In certain cases to remedy this matter overtaking sections can be provided. However, this assumes there is a central process control system to inform the truck how to bypass the queue.
- [ ] On highly frequented main lanes care should be taken to ensure that on turning into a secondary lane the first block section in the secondary lane is kept as short as possible so that the main lane quickly can be made free again. In certain cases the main lane can be relieved through secondary lanes which join in the central station.
- [ ] Load transfer stations on highly frequented sections should always be situated on additional secondary sections. Reversing pallet trucks should be avoided since they can prevent other trucks from moving forwards.
- [ ] A priority ruling at junctions with the main lane ensures that the through traffic has right of way.

Truck selection and load transfer. The choice of truck should be made with close consideration of the transport function, the material flow densities and the overall process organisation.

First, it has to be established if tractor train operations are justified bearing in mind material flow between the individual load transfer points or whether individual trucks offer a better solution.

If the transport order usually requires one load unit then the single truck solution is best since in a tractor train the assembly of trailers in the correct order for the destination points as well as the disconnection of trailers at the individual load transfer points leads to additional handling.

The next step is a decision on whether load units should be set down and picked up at floor level, or whether load transfer stations should be used for handling the loads. Although use of load transfer stations inhibits system flexibility it has the advantage that space requirements are considerably less for large throughput performances. The throughput can be increased with additional units so the whole process has a much tighter organisation.

Last, but not least, the choice of load transfer depends on the existing physical interfaces; sometimes it has to be adapted to local conditions at the various destinations. A systematic approach to the selection of suitable load transfer options is shown in Fig. 44.

The final definition of the type of truck can be made using the basic specifications of the different types of truck described in Chapter 3. The key factors here are the truck's steering system, the truck length and height, available battery capacity and the facility to install special handling equipment.

Battery capacity. Electrical drive systems are durable and require little maintenance; in addition battery operated industrial trucks have little effect on the environment. However there are certain disadvantages associated with load battery weight, operating range and the expense involved in a suitable system for recharging the batteries (see Chapter 5).

The capacity of batteries and the charging technology must be matched so the truck remains operational under continuous conditions. On the other hand the battery should not have any surplus capacity in the form of unused dead load accompanying the truck. When deciding battery capacity the maintenance system for the batteries must also be taken into consideration.

In this respect it is desirable to have automatic systems for topping up batteries with water and connecting to the mains. Both techniques make it possible to reduce the number of personnel necessary during recharging. Up till now two systems have been used for topping up with water (33):

☐ Aquamatik system is an automatic method in which floating plugs are used to control water level.

☐ Aquagen system is a technique in which the conventional inspection plugs are replaced by regenerators.

As a result of these measures the maintenance intervals of normal single-shift operation can be extended to several weeks.

The effect of automatic topping up is enhanced if the batteries can also be connected automatically to the charging units. One of three basic methods can be used here.

When the battery capacity drops to 20% of the rated capacity or upon completing the last transport order the trucks halt at the charging points; each charging point has a stationary transfer unit which is coupled to the truck. This technique is used especially with compact skid tractors.

The second method incorporates the automatic charging system on board the truck. In this case the charging units are connected to a contact plate at floor level, so that when the AGVS truck is positioned exactly over this plate for charging an arm moves down to make contact. This arm first cleans the plate with a brush and then makes the contact through spring-loaded bolts. After an exchange of signals the charging current is only then switched on by the truck (33). The connection to the electricity supply can also be made through an overhead electric rail which the truck reaches with a mast.

The third method is similar to the previous methods but makes use of pauses during the production cycle to recharge the batteries. This option is especially well suited when AGVS trucks are integrated into the production cycle as fitting or work platforms and have to wait for certain production processes to be completed. Providing such an intermediate charging operation does not lead to any release of gas pressure it does not harm the battery and does not have to be evaluated as a complete charging cycle. When choosing battery capacity the proportion of possible intermediate charging, insofar as it is provided in the production process, should be considered. The calculation of the required battery capacity can then be carried out as follows.

For example; when servicing a high-bay store the AGVS suppliers propose 24 volt/210 Ah PzS batteries for the trucks. It needs to be checked if this battery capacity is adequate for single-shift operation under the conditions that: – 80% of the total battery capacity can be used (or that they can be discharged up to 20%); – the efficiency of the electric motor, (or the ratio of the output power to the consumed power) is 50%; and that – approximately 60% of the shift time is used by the trucks for travelling.

Power for traction motor accounts for the current consumption; the lift motor is used only rarely and then for seven seconds. Thus, the average power of the traction motor should be taken as 0.5 kW.

In the calculation procedure equations (9), (10) and (11) apply to the calculation.

$$(9) \quad \eta_{EM} = \frac{P_{output}}{P_{consumed}}$$

where η_{EM} is the eficiency of the electric motor
P_{output} is the average power output by the electric motor
$P_{consumed}$ is the average power consumed by the electric motor

$$(10) \quad Batcap_{avail} = \frac{Bat_v \cdot Bat_{Ah} \cdot \gamma}{1000} \text{ kWh}$$

where $Batcap_{avail}$ is the available battery capacity in kWh
Bat_v is the battery voltage
B_{Ah} is the number of ampere hours of the battery

γ is a factor for accounting for the permissible charging of a battery without damaging the battery = 0.8 i.e. up to 20% of the total battery capacity can be discharged.

(11) $Batcap_{req} = t \cdot P_{consumed}$ kWh

where $Batcap_{req}$ is the required battery capacity in kWh

t is the travel time during the operating time without recharging in hours.

The following result was obtained for this example. From equation (9) the consumed power was calculated as 1kW; and the available battery capacity is then 4.032 kWh from equation (10); with a travel time of t = 4.8 hours the required battery capacity from equation (11) is 4.8 kWh. The battery capacity proposed by the manufacturer is therefore too small and should be made larger.

Determining the number of trucks. The number of trucks for a driverless transport system is calculated from:

☐ The number of load units which must be transported between sources of material flow and destinations of material flow (quantity movements per transport link with the respective distance and with respect to a time unit)
☐ Number of load units which can be picked up by the truck in one transport operation.
☐ Number or proportion of return trips can be made.
☐ Speed of the trucks.
☐ Load transfer cycle times.
☐ Cycle times for additional systems such as lift journeys, passages through gates and so on.
☐ Manipulation and switching times during the transport operation.
☐ Waiting times at the block points.
☐ Availability of the trucks or of the entire AGVS.

It is evident that the most varied stochastic influences which lead to a large number of unplanned events result in local waiting times and queues and cannot be accounted for by a simple deterministic calcuation. This means that in a complex system the number of trucks can be determined only by performing a correct simulation, even if this involves considerable expense.

However, if in the case of relatively simple systems a rough calculation is considered adequate the following procedure can be adopted.

First, all transport orders have to be sorted in a from/to matrix for the period with peak demand and the number of load units to be transported established (Fig. 85). If the number of load units which can be assigned to one truck per transport operation and the possibility of return trips are known then the number of load units to be transported can be converted to the number of required transport operations in the from/to matrix.

The necessary cycle time is then calculated for each transport operation, that is the total travel time using distances and truck speeds but taking rough estimates of waiting and unproductive time. The summation of all transport cycle times for the hour of peak demand as well as the availability of the trucks or of the entire AGVS makes it possible to calculate the number of trucks from the equation:

$$(12) \quad Z_{Truck} = \frac{\sum_{i=1}^{n} t_i \cdot h_i}{3600 \, \eta}$$

where Z_{Truck} is the number of trucks necessary
t_i is the cycle time of a transport operation for the transport connection i from the set of all transport connections, $i = 1 \ldots n$, in seconds.
h_i is the number of transport operations to be carried out for the connection i for the hour of peak demand
η is the availability factor for the trucks or for the entire DTS.

The necessity of simulation. A number of questions arise with the operation and planning of complex transport systems, such as the limiting capacity or the throughput time of transport operations. These cannot be solved simply with mathematical analytical procedures or the personal experience of the designer, since here the behaviour of the system over a period of time has been left out of the considerations.

Such transport systems are usually characterised by two criteria: The system exhibits dynamic behaviour; that is, the state at a point in time t_2 is directly dependent on a previous point in time t_1. And secondly, the system is subject to stochastic influences; that is, random variables such as transport order structure which cannot be predicted.

Simulation is a suitable aid for representing and indicating the desired performance and economic viability of a planned material flow system (48). The main advantage is that a given or planned system can be tested or designed without having to interfere with the actual process. Useful simulations can be carried out only with the help of computers using suitable simulation languages. According to the VDI recommendation 3633 (draft) simulation is defined as: '... the imitation of a dynamic process in a model in order to reach conclusions which can be transferred to reality'.

The use of simulation techniques do not replace planning effort. On the contrary it requires systems planning as a first draft in order to construct the model. As a heuristic procedure (a procedure which does not consider the sum of all conceivable constellations) simulation is not an optimising procedure. However each simulation run provides information on the system behaviour in response to specific questions. The performance of several simulation runs makes it possible to recognise trends and to reveal areas for improvement. Only with an iterative procedure is it possible to reach the 'best possible' system configuration (48).

The main effort involved in simulations is to design the model. The quality of the model and its closeness to reality determine the quality of the simulation results. An adequate representation of reality in the form of a model demands the following detailed information from the planner: the data base structure, the performance structure, and the organisation and process structure including their interfaces.

The effort involved in building the model must sometimes be limited by provisionally restricting its scope to critical sub-areas within the system. Although reducing the required effort such an inexact model has only a restricted power of prediction. The creation of a simulation model involves on the one hand the conversion of the actual system into a mathematical-logical description and on the other hand the conversion of the mathematical-logical model into a form suitable for computer input. The model can be formalised using standard problem-oriented languages or by special simulation languages which are more user-friendly.

With the help of interactive simulation (49) it is possible to organise the structure and make alterations to the data base independently of the simulation process, as well as to vary certain parameters during the simulation run. In this way current events can be displayed on a colour-graphic VDU as they occur.

The application area for simulation within the framework of material flow planning involves existing material flow systems as well as designing planned material flow systems. According to (48) the following questions can be answered with the help of simulation.

☐ Verification of function and performance.
☐ Analysis of system behaviour under different operating requirements.
☐ Effect of failures of individual components.
☐ Dimensioning of transport equipment and ancillary equipment.
☐ Utilisation of transport system components.
☐ Dimensioning of intermediate buffers.
☐ Design of transport paths.
☐ Representation of interfacing problems.

When planning new material flow systems simulation can perform the following functions:

☐ Comparison of the dynamic system behaviour for alternative design options as the basis for system selection; that is a comparison between the use of AGVS tractors on the one hand or single trucks on the other.
☐ Verification of the performance and functioning capacity of a planned material flow system in order to reveal complex system relationships; as well as for determining the limiting capacity, the average throughput times per transport operation and the formation of queues. Instead of determining the limiting capacity from a given number of trucks, the number of trucks can be calculated also from the given demand; as well as the

limiting capacity determined when additional trucks are introduced or the number of trucks is reduced.
- ☐ Iterative application of simulation in the planning process as a way of improving the system in stages so that local bottlenecks within the network are eliminated.
- ☐ Analysis of possible faults with the aim of developing strategies for emergency operation.
- ☐ Analysis of starting behaviour.
- ☐ Determination of system and strategy errors during the control process.

The necessity for exploiting the multi-faceted possibilities of simulation results from the existing complexity of the transport system and the subjective assessment of the planner who has to weigh up the effort involved in the simulation relative to the expected results. It should be taken into consideration here that the simulation does not replace the creative activity of the planner but is an aid to improve his design. The quality of the preliminary planning work makes it much easier to carry out the simulation, as shown in the flow diagram, Fig. 87.

Within the framework of planning new systems the joint participation in the overall cost of the simulation by the system manufacturer and the potential user has proved an effective solution. By sharing the cost with both partners involved in the project a more accurate and reliable simulation can be produced, at least as far as the systems analysis and data preparation are concerned. The joint undertaking of the simulation does not exclude contracting a third party to implement the simulation including the evaluation of the results.

Linking AGVS to a high-level process computer. Complex systems involving process computers require in the first instance the careful compilation of a performance specification in order to prevent inadequacies from appearing once the contract has been awarded. This is important from both the user's and the supplier's points of view (34, 56).

These days, planning the organisation of an automated materials flow system is often started only when the overall building and installation design has been established. Therefore it is required that the scheduling and organisation should proceed step-by-step in parallel with the planning of the material flow so that the two planning processes go ahead hand-in-hand (53). The organisation of the process computer operation, which is established as a result of this procedure, should lay down the performance specifications in a technically and legally binding form the system requirements for the tender or the supply contract.

Parallel planning improves the compatibility and the dovetailing of the design concepts for the material flows and the operating processes as well as the overall information processing. Practical experience has shown that these two planning processes should be staggered in time, that is they should proceed in such a way that after an analysis of the existing organisation a

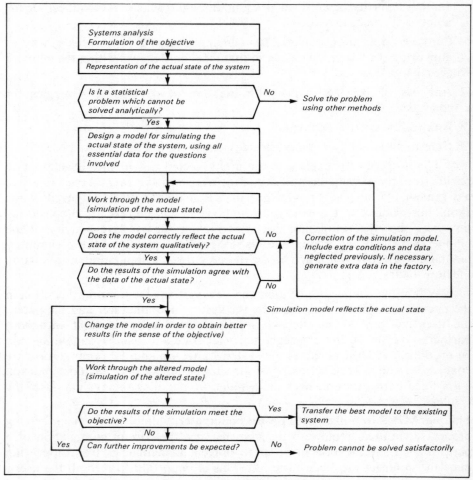

Fig. 87: Flow diagram for carrying out a simulation study

certain waiting period should be inserted in order to give material flow planners the opportunity to develop a rough design concept.

The essential content of such a performance specification is briefly described as follows (53):

Content of a performance specification

1. Description of the process

    Introduction to the problem, basic principles of the project, overall description of the required condition and possibly the actual condition; definition of the technical process.

2. System overview

    Analysis of the project and objectives; division of the overall problem, into a hierarchy consisting of system, sub-system and functions with corresponding description; specification of the performance data and the quantity data for the overall system; operating conditions and fault behaviour as well as other system requirements.

3. Technological functions

The technological functions should be clarified between the user and the supplier and laid down in detail and in a binding manner; interface descriptions and definition of input/output data should also be included.

4. Storage functions

The storage capacity includes the technological functions demand. This concerns its states as well as object descriptions (master data).

5. Human input/output

All interfaces between operators or users and the process data processing system.

6. Process input/output

Description of all interfaces between equipment (including other computers in the technical process) and the process data processing system.

7. Process control

Analysis of the time and logical relationships between the technological functions, including the production of a phase schedule for an entire working day.

9. Necessary operating media

Evidence of technical feasibility by defining critical system components; derivation of necessary operating media and other prerequisites (standard software, user software, hardware configuration).

9. Project planning

Planning the implementation and the responsibilities through scheduling, project organisation, acceptance procedures and customer supplies.

Typical weak points in performance specifications are the result of unsystematic description, a relatively inaccurate definition of the system and an inadequate description of the function. From previous experience in the implementation of automated material flow systems it can be assumed that the expenditure on the entire performance specification amounts to approximately 10% of the total software costs.

The compilation of the performance specification marks the end of the first phase in generating a computer system. The next two phases are: system implementation; and commissioning/bringing into operation.

The implementation phase for a computer design concept has few direct points of contact with the other components involved in setting up a material flow system. On the one hand it refers to the definition of the objective, as laid down in the performance specification, and on the other hand it is terminated by the works acceptance.

Nevertheless the implementation phase of the computer design concept is of fundamental significance because the structure of the system is laid down here and any subsequent alterations can be made only at great expense and at the expense of considerable delay (26).

The quality of the implementation of a computer design concept has a decisive effect on the success of the project so that any weak points must be picked up and eliminated at this stage.

Inadequacies in the implementation of the computer design concept result in the first instance from a poor performance specification, as discussed previously. In addition, care should be taken to ensure that the programming is clear and has been developed in stages. This means that even under severe time pressure the programming should be carried out only after completing the software design. Commissioning can be started only on site when the computer design concept has been completed and the works acceptance has been carried out.

In addition the interfaces with neighbouring systems should be considered also here. These should be made as simple, compatible and susceptible to simulation as possible in order to carry out commissioning in successive stages corresponding to the computer hierarchy.

Defects in the computer implementation can be identified in the physical material flow system; for example, software operating times which produce such long reaction times that queues build up in the transport process. Typical defects are: waiting for load units at the entry point, and waiting of trucks for reference values.

Ambiguous and complex material flow strategies can cause transport operations to accumulate at certain points, or enable priority assignments for certain transport operations to block the performance of other functions. The operating capacity of the overall system must therefore be taken as the key criterion when implementing the computer system (26).

The best possible operating capacity of the entire system is reached when the control design concept corresponds to the hierarchical requirements, an example of which is given in Fig. 52. In addition the system should be designed with a view to: availability, ease of expansion, behaviour over a period of time, and user-friendly peripherals.

Without proper project organisation and competition it is impossible to avoid inadequacies in complex material flow systems.

Tables 3 and 4 briefly describe the project phases of software management and hardware management.

	Phase	Content	Documentation
Documentation accompanying the project	Production of performance specification	• Specification of material flow • Specification of organisation 　– organisational procedures 　– dialogues 　– records/lists 　– emergency organisation 　– start/restart behaviour 　– interface specification • Specification of quantity/time diagram • Specification of project organisation 　– User participation 　– Deadlines/scheduling 　– Documentation standards 　– Acceptance procedures	Performance spec. Performance spec.
	Software design	• System design • File design • Preliminary module design • Integration of contents • Dialogue and listen formats • Specification of test strategies involving existing software modules	User manual System manual
	Programming	• Final module design • Coding • Module testing	Programming manual
	Integration	• Stepwise integration and testing of the total software system with simulated interfaces	
	Works acceptance	• Demonstration and acceptance of individual functions • Acceptance of the overall design with simulated interfaces	Acceptance report
	System test	• Testing with adjacent systems under operating conditions • Training under operating conditions	Revised user manual System manual
	Commissioning	• Test of the overall system under operating conditions	
	Function acceptance Trial run/performance test Operation under production conditions	• Acceptance of individual functions • Acceptance of the overall system in trial operation with performance test • Production support • Modification based on operating experience	Acceptance report Revised overall documentation

Table 3: Project phases in producing the software

Phase	Content	Documentation

Production of performance specification	● Specification of hardware configuration
	   – Memory capacities
	   – Suitable peripherals
	   – Spare parts, standby equipment
	● Specification of hardware interfaces to adjacent systems
	   (Computer link-ups, test point lists)
	● Specification of project organisation
	   – Deadlines/scheduling
	   – Training at manufacturer's works
	   – Acceptance procedures
		Performance spec.

Installation planning	● Installation planning
	   – Building measurements
	   – Power supply
	   – Environment
	● Cabling planning, observing relevant regulations
		Installation plan

Ordering	● Purchase technical support	
Installation test bay	● Installation monitoring ● Connection/integration special peripherals and simulation hardware ● System generation ● Reception and acceptance of supplies ● Operating media acquisition/management ● Test operation organisation	Hardware, system software, documentation (from the manfacturer) Generating report
Acceptance test bay	● Demonstration/performance testing with representative peripherals and simulated interfaces	Acceptance report
Assembly	● Verification of assembly requirements ● Supervision of dismantling test bay and installation on site ● Cabling testing ● Installation and connection to peripherals	Revised cabling plans and test point lists
System test/commissioning	● Commissioning of the hardware components in stages ● Training of the operating staff	
Acceptance	● Monitoring the scope of the supply Function acceptance within the framework of software acceptance	Acceptance report
Maintenance	● Organisation of maintenance	Log book

Table 4: Project phases in producing the hardware

Chapter Seven

Economic viability as a decision aid in selection

THE CHOICE of a transport system is, in the first instance, a technical problem, as described in the previous chapters. However, every decision involving the use of a transport system requires comparative economic studies (8, p. 316).

So a distinction should be made between approaches dealing with the economic viability of different system options when planning new installations and procedures for determining the economic viability during reorganisation or replacement investments. In the latter case, apart from the decision as to whether to replace an existing transport system and with what, there is also the decision at what point in time this is best done. In the case of planning a new system that point in time is fixed as a matter of principle.

Heinen (6, p682) makes a distinction in the types of investment calculation between sub-goal oriented procedures and end-goal oriented procedures. Sub-goal oriented procedures, (Fig. 88) are directed to short-term predictions whereas end-goal procedures are based on a total consideration of the investment, that is they consider the consequences of investment options right up to the end of their working life.

When selecting the investment calculation procedure for evaluating transport system options it should be borne in mind that transport systems provide assistance only in achieving the actual production performance of the organisation; in other words the application of a transport system has no actual market value.

In this respect procedures, such as the profit comparison method, cannot be used. Hartmann in particular makes this point in the evaluation of information systems which also have internal service functions; it should be noticed here that although the determination of the costs does not cause any great difficulty there is no performance in terms of value which can be exactly expressed in figures that can be set against them (4, p371).

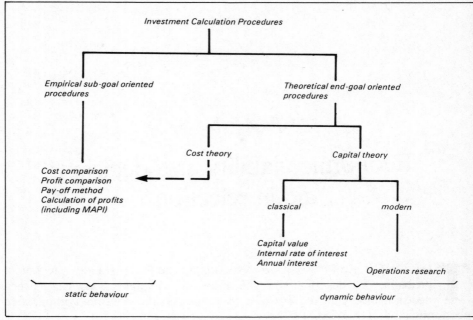

Fig. 88: Investment calculation procedures

In principle the known investment calculation procedures can be applied to investments in the service area if the profits appear as savings; these will be essentially labour savings (7, p383). Other savings, such as the reduction in down-time in production due to lack of transport availability, usually can be assessed only with difficulty. Therefore the known investment calculation procedures can be used in a limited fashion.

As a dynamic procedure Hartmann suggests in (5, p69) conversion of the investment formula:

$$(13) \quad \text{Comparison factor}_1 = \frac{\text{Cash value of the returns}}{\text{Cash value of the costs} + \text{initial outlay}}$$

to the following formula (14)

$$(14) \quad \text{Comparison factor}_2 = \frac{\text{Cash value of savings}}{\text{Cash value of extra costs compared to the old system} + \text{difference in initial outlay}}$$

which, as a quotient, sets the cash value difference of the extra costs and the cash value differences of the initial outlay against the savings.

If the comparison factor is unity then the profitability of the investment corresponds to the accounted interest which was considered in the cash value of the costs. The larger the comparison factor the more interesting is the investment.

Mellerowicz bases his approach for the economic selection of a transport system on the profit comparison calculation and suggests that the best transport system will have the maximum margin between return and costs (8, p316). According to this:

(15) $\quad F + P < (A + Gk) \cdot x + R$

where F is the fixed cost
P is the operating cost for the transport system relating to employment
A is the saving in labour costs
Gk is the other overhead costs
x is the (annual, monthly) duration of employment
R is an additional rationalisation profit

The calculation procedures based on equations (14) and (15) set the costs against profit or savings in the same way and, in contrast to the pure cost comparison calculation, consider the expected 'return' on the investment which in equation (14) is expressed in the form of profitability and in (15) in the form of an absolute amount.

Despite the difficulties associated with the determination of the 'return' within the framework of an investment for a transport system these procedures can be regarded as a basis for decision since they accurately reflect the individual operating situation. The viability of a replacement investment is then obtained from the difference amounts in comparison to an existing transport system.

The problem is different when planning a new system. In this case the relative advantages of a transport system can be calculated in comparison to another option. In this way the comparison between the planned system and the actual system of equations (14) or (15) is replaced by the comparison between one planned system and another (a plan-to-plan comparison).

However since the 'profit situation' when planning a new installation does not have to be measured in terms of an imaginary operation, the pure cost comparison calculation often can be considered to be adequate, especially since the plan-to-plan comparison is further burdened by a future uncertainty of the relative savings. On the other hand, running costs comprising depreciation on the initial outlay, the calculated interest in capital, the maintenance and servicing costs, as well as the necessary personnel requirements and other variable costs can be relatively reliably determined.

Maynard (7, p577) considers the cost comparison calculation as a suitable selection procedure for investment decisions in the material flow sector, irrespective of whether the investment involves a replacement or a new system.

The running costs of a transport system can be used as an aid for assessing transport systems for two reasons. First, the cost structure can be determined relatively easily; and secondly the cost comparison calculation, as a planning aid, allows a narrowing down in terms of the detailed technical planning of a transport system if first a test is made from the economic point of view (cost comparison calculation) to determine which systems are feasible for the case in hand.

It is obvious that while an economic comparison is an important criterion when making an investment other factors have also to be considered.

In the first instance these involve advantages which cannot be quantified or only with difficulty, such as improved scheduling or organisational simplification. Furthermore the technical system characteristics of transport systems have to be considered, such as the reliability, availability and flexibility of the system. A basis for an efficiency analysis for weighting the key variables in making investment decisions can be found in (10).

Objectives of cost analysis. Within the framework of a comparison of operating costs, costs which are incurred per transported unit when using different transport systems have to be determined. The purpose of the cost comparison, using date relevant to actual practice, is to establish which transport systems are best for which transport operations from the economic point of view.

The analysis should be considered as an aid in making a preliminary selection; it does not replace an investment calculation in a specific application. This cannot be achieved for the reason that when making a general abstract cost comparison it has to be based on average cost values which can deviate from a specific application one way or another.

Consequently, for defined transport operations the cost amount was not determined absolutely in Deutsch Mark (DM) but was established in terms of a rank order, based on comparative values expressed in terms of a characteristic value (DM per transported unit).

Furthermore, in terms of a sensitivity analyis it is shown how the costs per transported unit and of the transport system can vary is shown when the basic starting parameters, namely the transported quantity or the transport distance, are varied.

The importance of the effects of loading the transport system on the transport costs can be assessed by varying a number of parameters, for example by increasing the quantity to be transported in steps for a given distance. In order to display this cost factor continuously the cost values were derived by computer and drawn as an output by a plotter.

By making a sensitivity analysis the cost curves show any trends in the costs per transported unit and transport system costs when varying the parameters; in this way a prediction can be made whether a planned decision remains stable in terms of its outcome when hypothetical changes are made to the performance.

Definition of some parameters. Within the framework of this cost comparison a transport operation is defined as picking up a load at point A, and transporting it to the point B with load transfer at that point. In this way the necessary organisational and technical effort can be determined per transport operation. A transport operation defined in this way does not consider any return

ECONOMIC VIABILITY AS A DECISION AID IN SELECTION

journeys, thus a transport cycle always involves an empty journey as the return trip to the starting point.

The transport operation is to be carried out within a two-shift operating day whereby the average number of operating hours per year is calculated to be 3,840.

The goods to be transported are standard internal items carried on a load unit. The load is 500kg. The load handling equipment is a Europallet measuring 1,200 x 800mm.

The minimum relevant internal transport distance is 25m. By doubling this value a series of transport distances can be built up: 50, 100, 200, 400, 800, and 1600 with 3200m as the maximum transport distance.

By defining the hourly mass flows the entire spectrum of transport capacities in standard installations can be covered. As with the definition of the transport distances a minimum transport quantity of four load units per hour is first considered. By doubling this value the series 4, 8, 16, 32, 64, 128, 256 up to a maximum flow of 512 load units per hour can be defined.

From the large number of internal transport systems on the market capable of handling one-piece goods the following competing systems were selected for the operating cost comparison:

- Hand pallet trucks
- Reach mast trucks
- Electric tractors with 4 trailers
- AGVS tractors with 4 trailers
- AGVS pallet trucks
- Monorail conveyors
- Underfloor chain conveyors
- Power-and-free conveyors
- Powered-roller conveyors

In order to carry out the cost comparison the following assumptions also had to be made.

- Investment and operating costs for a central process computer are not included for systems suitable for automation for reasons of the comparability with transport systems with a low level of technology.
- When determining the investment costs for fixed installations the special construction costs involved in building the structures are not included. This applies especially to the monorail conveyor, the power-and-free conveyor and the underfloor chain conveyor. The assembly costs of the system components are considered in the investment costs of the installation for the individual transport systems insofar as this is necessary.
- With the hand pallet trucks, in contrast to all other transport systems, no reserve transport capacity in terms of time is allowed for, thus the hand

pallet truck is only occupied and engaged if a transport operation is in hand. This approach is justified to the extent that the hand pallet truck is a means of transport with a low value and as a rule is used only for spontaneous transport needs. In industrial practice this means that for example a member of staff in the stores or in production only uses the hand pallet truck for occasional transport operations which take up a small proportion of his actual main activity. With all other transport systems, insofar as this is necessary for the organisation, personnel capacity is maintained over the two-shift period.

Cost accounting data, fixed costs and variable costs. The operation of transport systems results in fixed and variable costs. Fixed costs are incurred independently of the degree of loading whereas variable costs depend on the degree of loading. The fixed costs cannot always be separated exactly from the variable costs, thus the allocation depends amongst other factors on the subjective assessment of the cost accountant and on the existing possibilities of the cost analysis. Within the framework of this cost comparison the following were defined as fixed costs:

- Depreciation. The calculation of the annual depreciation is carried out using the straight-line method of depreciation with equal annual contributions.
- Cost accounting interest. The amount of the interest on the invested capital is calculated by the average interest method whereby interest is accrued on capital which is tied up on average over the whole service life. In conjunction with straight-line depreciation half of the material assets should be used for the averaging of interest which include also the assembly costs for the system. The interest rate was fixed at 10%.
- Factory costs. Insofar as production areas or other useful areas are permanently occupied by transport systems the associated costs must be taken into consideration. Such costs include the costs for air-conditioning, light and so on.
- Maintenance costs. The running, servicing and repair costs are combined together to give the maintenance costs. Depending on the system they were taken as a percentage of the overall investment.

The variable costs to be determined are assigned to the following types of cost.

- Personnel costs. The personnel costs involve the direct costs which the employer pays the employees as salaries. In addition there are also the indirect costs which the employer incurs as the result of social legislation. Considering the legal protection of the personnel the question arises as to what extent the personnel costs can be considered as variable costs; that is it should be established whether by stricter application of the 'variable' costs concept the possibility exists of making personnel costs directly proportional to any fall-off in employment. This approach might justify assigning personnel costs to the fixed costs at least in the medium-term.

However within the framework of this comparison what is of interest is the personnel complement which is necessary for precisely defined transport operations. Fluctuations in loading in respect to a transport system should be evaluated in terms of their effect on personnel requirements so that for all previously defined debit items in the cost comparison the associated personnnel costs should be determined.

- ☐ Electrical charging costs. The electrical charging costs are calculated from the electricity consumption of the trucks, the cost per kWh for electricity supplied to the battery through the battery charger, and the charging factor, that is the efficiency of the charger.
- ☐ Battery costs. The battery costs can be considered as variable since after 1,500 discharges the batteries are exhausted and have to be replaced. Depending on the degree of utilisation of electric trucks recharging is necessary during the same working day or only after the second working day.
- ☐ Energy costs. All non-freely moving transport systems take electrical power for their drives from the local or internal power supply system. These energy costs are calculated from the power demand of the transport system, the duration of operation and the current electricity price.

A survey was made among users in order to determine current values for the different types of costs. A special data form was used for this. Insofar as there were not any systems available according to the specifications given in the transport quantity/transport distance matrix the manufacturers' data were obtained for the installation costs.

The computer-based calculation of the transport costs per transported unit for each transport operation was carried out for non-conveyor transport systems by determining the necessary number of transport units. According to the number of transport units which provide the necessary capacity for the transport operation the variable costs are assigned as being proportional.

With continuous systems such as chain conveyors, a limiting capacity is given independently of the transport distance by virtue of the continuity of the process, [equation (1)]. With mass flows below the limiting capacity, variable costs can sometimes be obtained by switching the system on and off. However as a general rule transport readiness should be maintained over the entire shift operation.

For most internal applications the limiting capacity of a continuous system is adequate. Only in special cases when the limiting capacity is exceeded does a second system have to be installed. As with non-continuous transport systems the variable cost due to depreciation is included in the continuous systems.

The entire cost calculation for all transport systems and all transport operations was based on the time unit of 1 hour. As a result the costs per transported unit were displayed with the help of a plotter with the transported load units per hour as the divisor for each transport operation.

Operating cost comparisons. The results of the cost comparison are displayed in the form of cost curves. Based on the parameters transport quantity (load units per hour) and distance the cost comparison curves are obtained for eight transport distances as reference parameters for all the transport systems under consideration, (Figs. 89–96). It should be noted that logarithmic scales are used on both axes.

The inverse representation is shown in Figs. 97–104. With reference to a given transport quantity the operating costs are displayed for all transport systems for different transport distances.

The curves show clearly the different cost behaviour between the conveyor and non-conveyor systems. The conveyor systems have high limiting capacities available with one installation (a roller conveyor system) and so with increasing loading exhibit constantly falling operating costs per transported unit – see Fig. 89.

The situation is different with the non-continuous systems. Thus for example in Fig. 89 the operating costs per transported unit for a mast reach truck fall as fast as the maximum loading of the vehicle, which is approximately 50 load units per hour. When this limiting capacity is reached a second vehicle must be acquired, this produces a sudden increase in fixed costs. Above this limiting capacity the operating costs per transported unit fall again until with complete utilisation of both vehicles the minimum cost factor is obtained.

The following information can be derived from the comparison of the transport systems.

Hand pallet trucks. Hand pallet trucks are suitable for short distances and low transport capacities. With a transport distance of 25m and with a transport capacity of more than 10 load units per hour other transport systems are already more economic.

Mast reach trucks. Mast reach trucks are the most expensive transport system over a wide range. Assuming that normally more than eight load units have to be transported, (Fig. 94) then in terms of operating costs mast reach trucks cannot compete with other transport systems. The fact that the mast reach truck is the most widespread transport system in use is because vertical transport (lifting and lowering) is often directly linked to horizontal transport at load transfer points. The mast reach truck is popular also because it is an extremely flexible form of transport and it is not necessary to organise the entire material flow systematically. These advantages are however often offset by relatively high costs.

Electric tractors (driver-operated). The driver-operated electric tractor is a means of transport which under the operating conditions assumed here (four trailers = eight pallets) lies in the medium cost range. Insofar as conveyor systems cannot be used more favourable operating costs can be obtained only with the help of AGVS tractors.

AGVS tractors. AGVS tractors are transport units with favourable operating costs which, at distances of 100m and a mass flow of 10 units per hour, exhibit lower operating costs than any other non-conveyor transport system, (Fig. 91). Only at very high capacities, that is at mass flows above 40 load units per hour do conveyor systems have distinct advantages.

AGVS pallet trucks. The operating costs of the AGVS pallet trucks lie clearly below the operating costs of the mast reach truck over the entire range of the investigation. Since in one transport trip only one pallet is transported they are suited for quite specific transport operations without a large throughput. In comparison to the driver-operated electric tractor the pallet truck is better in terms of cost until the electric tractor is at least half-loaded, (Fig. 93).

Monorail conveyors. As the cost comparison curves show the monorail conveyor is an extremely favourable form of transport. However, with this comparison it should not be forgotten that in the installation costs for the monorail conveyor the additional costs for erecting the overhead rail have not been included. This cost item cannot be evaluated as a standard value since the building costs involved vary considerably with the specific conditions.

Conveyor systems. With the exception of the powered roller conveyor, which involves considerable costs for the individual drive systems, the underfloor chain conveyors and the power and free conveyors are the best transport systems in terms of operating costs insofar as the prevailing operating conditions permit their use. The fall in costs is particularly noticeable at high transport capacities. The determination of the operating cost curves based purely on calculation ignores the fact that chain conveyors can be used only above distance of approximately 100m.

In order to make a critical interpretation of the operating cost curves the scaling of the cost factors has to be taken into consideration. This means amongst other things that:

☐ The required carrying capacity of the individual systems has a considerable effect on the investment costs for the transport system. This applies especially to the monorail conveyor which, with increasing carrying capacity, requires excessive investment.

☐ The branching of the network, the complexity of the installation or the number of destinations to be accessed have a considerable effect on the investment and on the operating costs. In the examples given here only the transport from position A to position B is investigated.

☐ The operating conditions have a significant effect on the amount of the maintenance and servicing costs.

For these reasons these cost comparison curves as already mentioned at the outset are only a guide in making a preliminary selection of transport systems.

AUTOMATED GUIDED VEHICLES

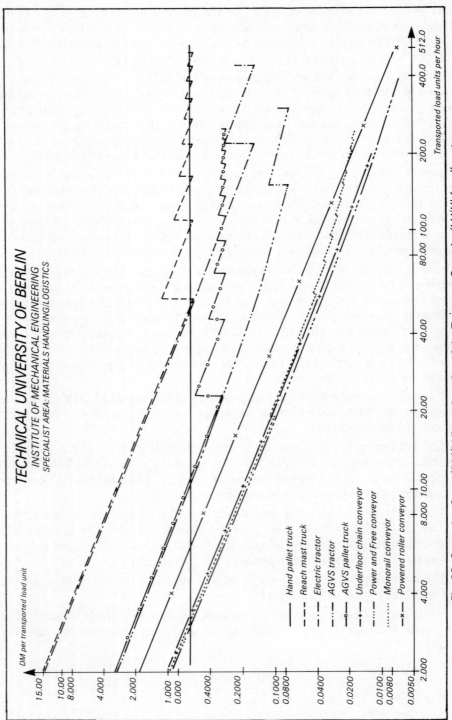

Fig. 89: Operating Costs (DM/LU) as a function of the Transport Quantity (LU/H) for all systems. Transport Distance is 25M and LU is Load Unit

ECONOMIC VIABILITY AS A DECISION AID IN SELECTION 141

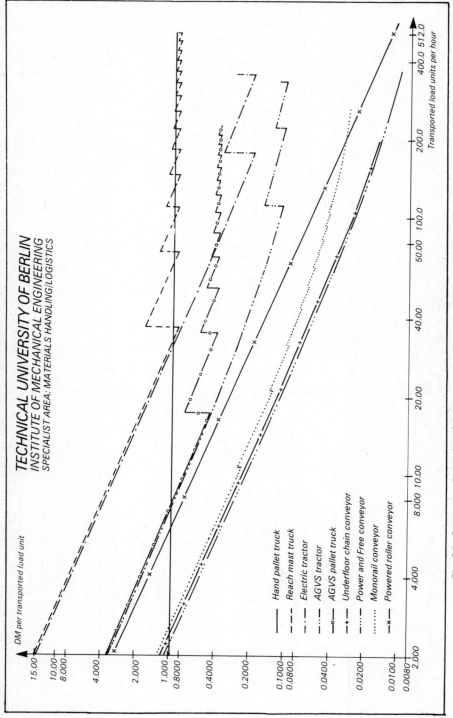

Fig. 90: Operating Costs (DM/LU) as a function of the Transport Quantity for all systems. Transport Distance is 50M

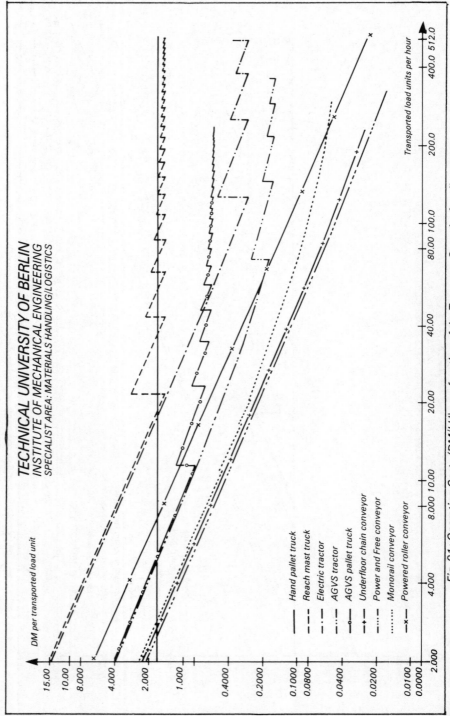

Fig. 91: Operating Costs (DM/LU) as a function of the Transport Quantity for all systems. Transport Distance is 100M

ECONOMIC VIABILITY AS A DECISION AID IN SELECTION

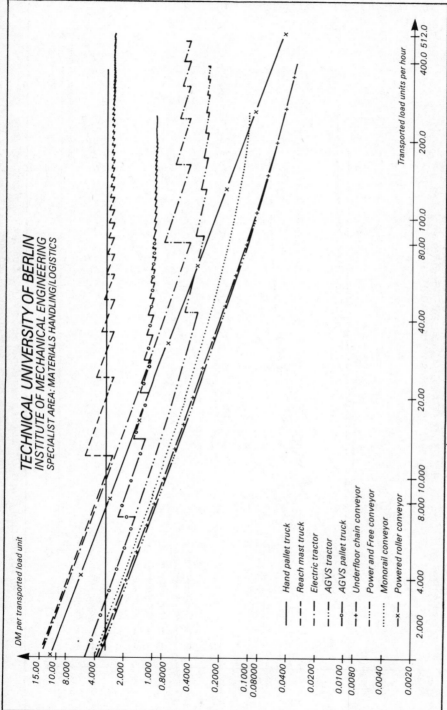

Fig. 92: Operating Costs (DM/LU) as a function of the Transport Quantity (LU/H) for all systems. Transport Distance is 200M and LU is Load Unit

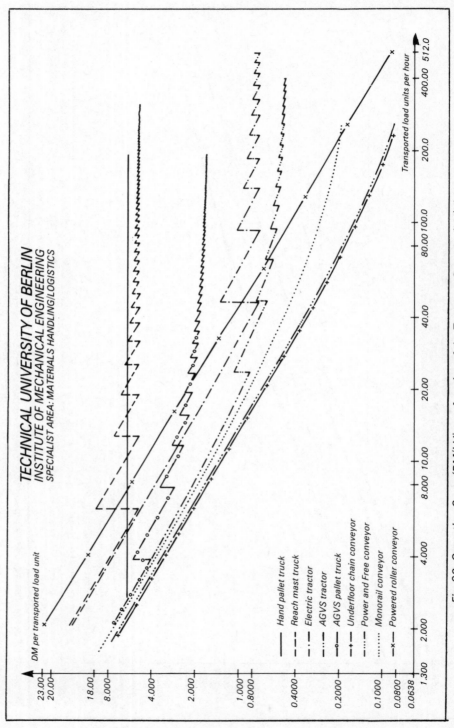

Fig. 93: Operating Costs (DM/LU) as a function of the Transport Quantity (LU/H) for all systems. Transport Distance is 400M and LU is Load Unit

ECONOMIC VIABILITY AS A DECISION AID IN SELECTION 145

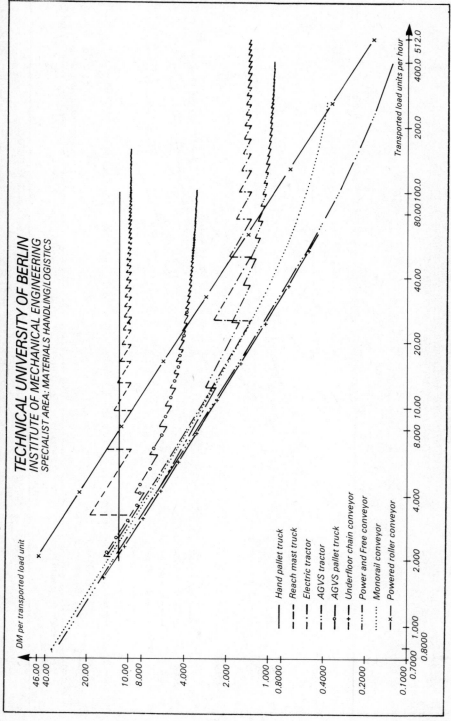

Fig. 94: Operating Costs (DM/LU) as a function of the Transport Quantity for all systems. Transport Distance is 800M

AUTOMATED GUIDED VEHICLES

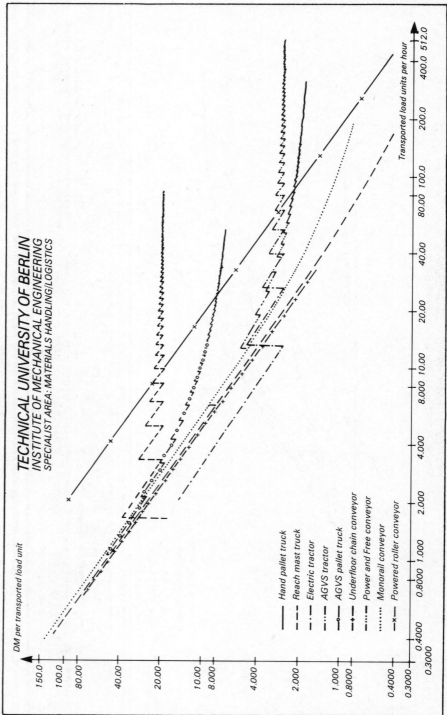

Fig. 95: Operating Costs (DM/LU) as a function of the Transport Quantity for all systems. Transport Distance is 1600M

ECONOMIC VIABILITY AS A DECISION AID IN SELECTION 147

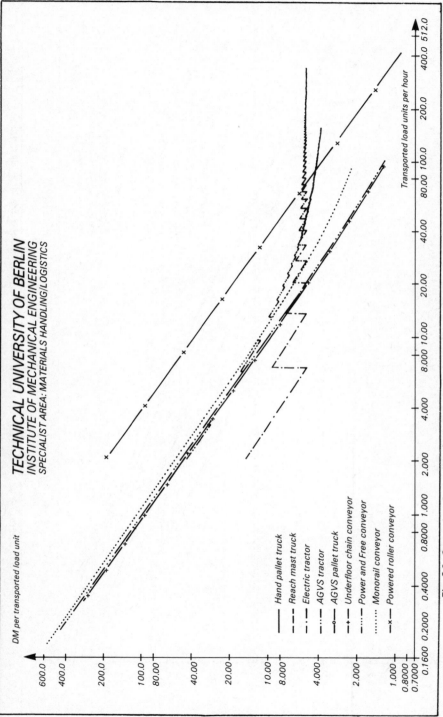

Fig. 96: Operating Costs (DM/LU) as a function of the Transport Quantity (LU/H) for all systems. Transport Distance is 3200M and LU is Load Unit

AUTOMATED GUIDED VEHICLES

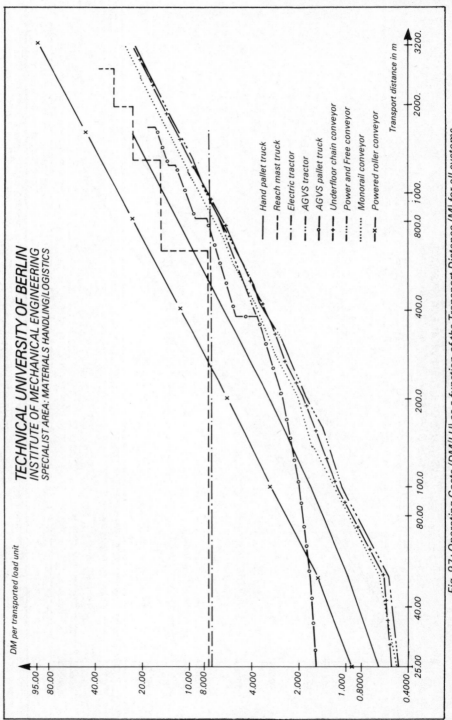

Fig. 97: Operating Costs (DM/LU) as a function of the Transport Distance (M) for all systems. Transport Quantity is 4 (LU/H)

ECONOMIC VIABILITY AS A DECISION AID IN SELECTION

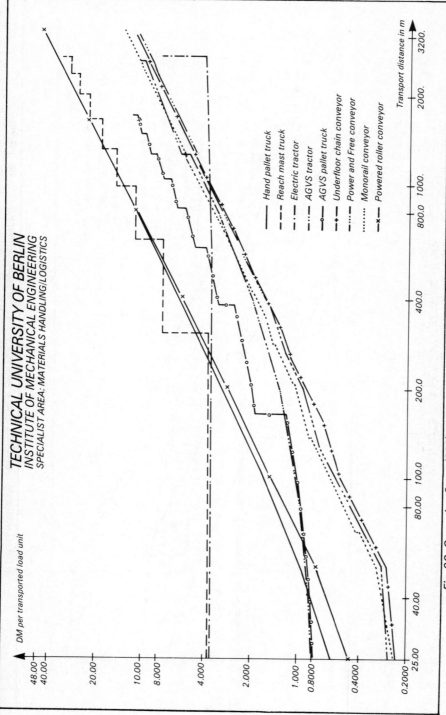

Fig. 98: Operating Costs (DM/LU) as a function of the Transport Distance (M) for all systems. Transport Quantity is 8 (LU/H)

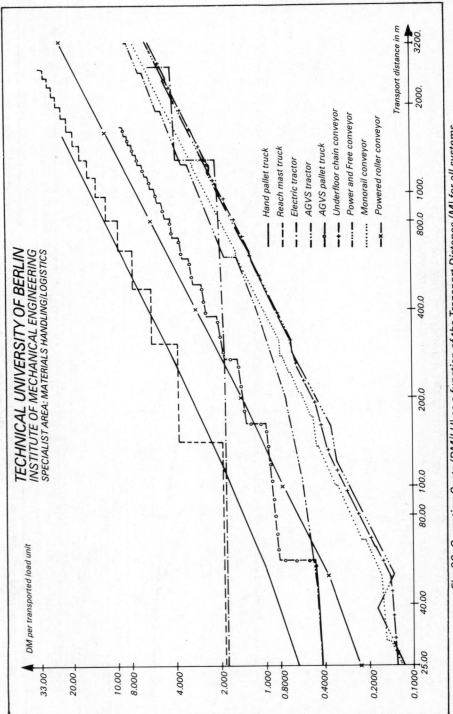

Fig. 99: Operating Costs (DM/LU) as a function of the Transport Distance (M) for all systems. Transport Quantity is 16 (LU/H)

ECONOMIC VIABILITY AS A DECISION AID IN SELECTION

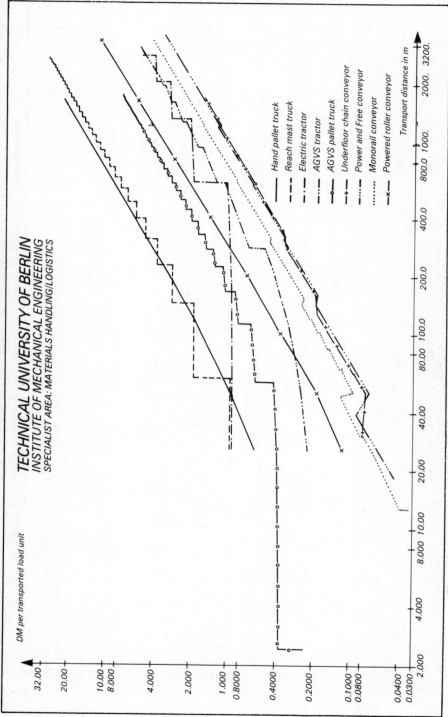

Fig. 100: Operating Costs (DM/LU) as a function of the Transport Distance (M) for all systems. Transport Quantity is 32 (LU/H).

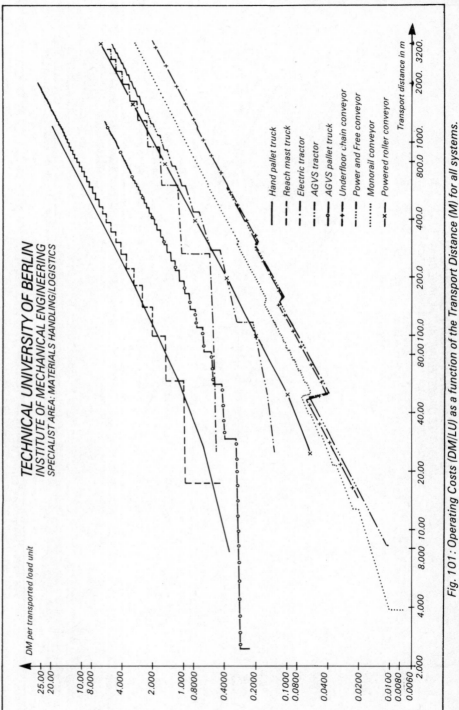

Fig. 101: Operating Costs (DM/LU) as a function of the Transport Distance (M) for all systems. Transport Quantity is 64 (LU/H)

ECONOMIC VIABILITY AS A DECISION AID IN SELECTION

153

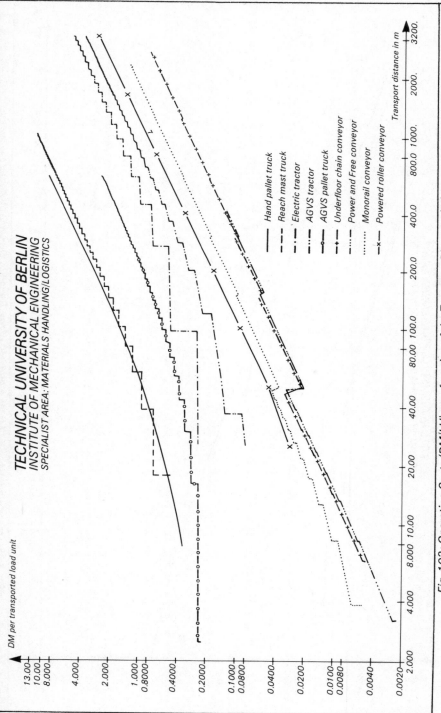

Fig. 102: Operating Costs (DM/LU) as a function of the Transport Distance (M) for all systems. Transport Quantity is 128 (LU/H)

154　AUTOMATED GUIDED VEHICLES

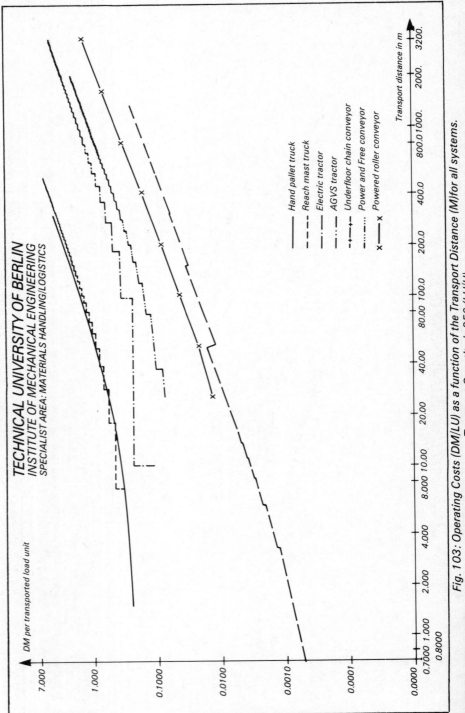

Fig. 103: Operating Costs (DM/LU) as a function of the Transport Distance (M) for all systems. Transport Quantity is 256 (LU/H)

ECONOMIC VIABILITY AS A DECISION AID IN SELECTION 155

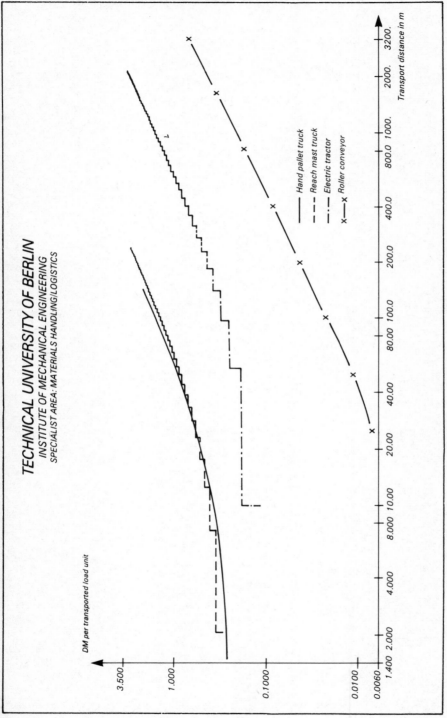

Fig. 104: Operating Costs (DM/LU) as a function of the Transport Distance (M) for all systems. Transport Quantity is 512 (LU/H)

Chapter Eight

Reasons for implementing an AGVS

WITHIN the framework of the market survey (22) which was carried out users of AGVSs were asked to give their reasons for implementing the equipment. The individual criteria were rated from 1–10 with 1 representing 'unimportant' and 10 'very important'. The five reasons for implementation were:
- Risk of damage during transit.
- Flexibility with respect to expansion, shut-down and reorganisation
- Automation of material flow
- Improved process organisation
- Cost advantages compared to other systems
- Savings in personnel

The evaluation showed (Fig. 105) that the four last-mentioned criteria are of special importance but also that the flexibility of the system, with a view to future developments, was attributed a relatively high importance.

Humanising assembly work. Insofar as AGVSs have been used the results of humanising assembly work by moving away from synchronised assembly line work have not yet been adequately evaluated from a statistical and ergonomics point of view to the author's knowledge. As a result no generally valid statements can be made.

The most important provisional results from a study (14) of the working methods of a Swedish enterprise are given below. The working system is essentially based on the principles:
- Assembly without synchronisation
- Combination of work content
- Group work with extensive self-determination
- Quality control within the group

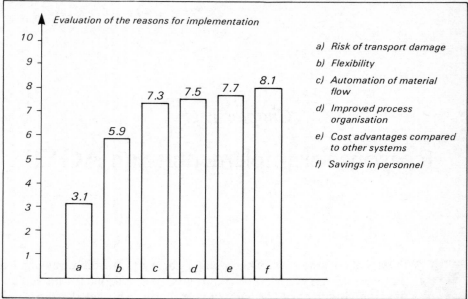

Fig. 105: Reasons for implementing AGVS from the user's viewpoint (arithmetic mean of the given evaluation grades 1–10; 1 = most unimportant; 10 = most important)

☐ Room for varying the working intensity

In comparison with conventional working systems it was found that:
☐ The time for assembling a vehicle remained the same (14, p36)
☐ Absenteeism from work due to illness and so on fell by 5.2%
☐ Fluctuations within the workforce fell by 4.5% (13, p37).

Compared with conventional assembly works these positive effects can be set against extra costs of 10% for the overall assembly plant.

Meanwhile, the Federal Ministry for Research and Technology in West Germany has published a further study in group work using driverless transport systems or AGVSs (2). In view of the different ideas which the participants had of the objectives of the project and the relatively short time that the project had been running before publication the results have only limited validity.

Economic viability. The economic viability of AGVS results from the overall benefit as seen from a business management point of view. This involves savings in costs, the non-quantifiable effects on the organisational sector and the technical performance capacity of the system.

These three areas can overlap in content. Each represents, with varying emphasis, a reason for implementing an AGVS within the framework of an investment decision. The investment decision on relatively complex transport system options demands that the decision-maker evaluates both the quantifiable and the non-quantifiable information.

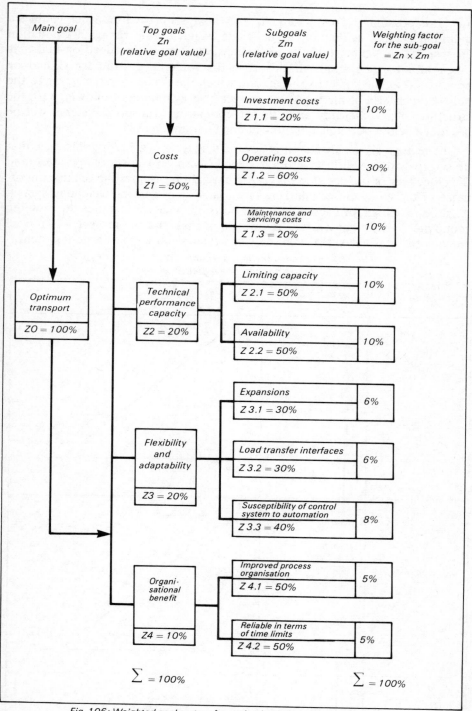

Fig. 106: Weighted goal system for evaluating internal transport systems

It is possible to conduct an eficiency analysis procedure which enables a multidimensional evaluation to be made using a wide range of evaluation criteria. At the same time the preference structure of the decision-maker is taken into consideration. In this case the preference structure should be understood as the subjective weighting which the decision-maker gives to the individual criteria of the transport project he is evaluating. Following (10) the structure of an efficiency analysis for comparing transport systems will be described below. A detailed treatment can be found in (13).

The analysis is carried out in several stages. A system of goals must first be set up with weighting factors assigned based on transport operation and the objectives of the decision maker. The weighting defines the relative importance of the goals to one another. The goal system and the weighting system must be redefined for each investment decision since for instance the required economic viability can vary as a function of time and the project − as indeed can all other variables. Fig. 106 shows an example of a possible goal system.

Fig. 107: Example for the determination of the target value for an economic evaluation analysis

SUB-TARGET	CHARACTERISTIC VALUE	EVALUATION
Z 1.1 Investment− costs	Investment− amount in DM	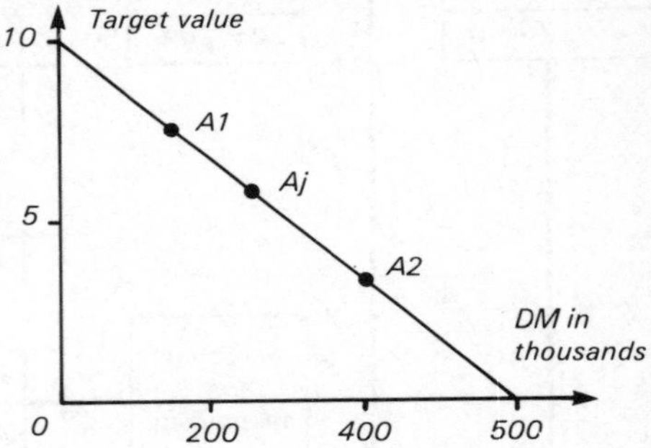 Target value plot: line from (0,10) to (500,0); points A1 ≈ (150, 8), Aj ≈ (250, 6.5), A2 ≈ (400, 3); x-axis: DM in thousands (0, 200, 400, 500); y-axis: 0, 5, 10
Z 1.2 Operating costs	Operating costs in DM, per transported loaded unit	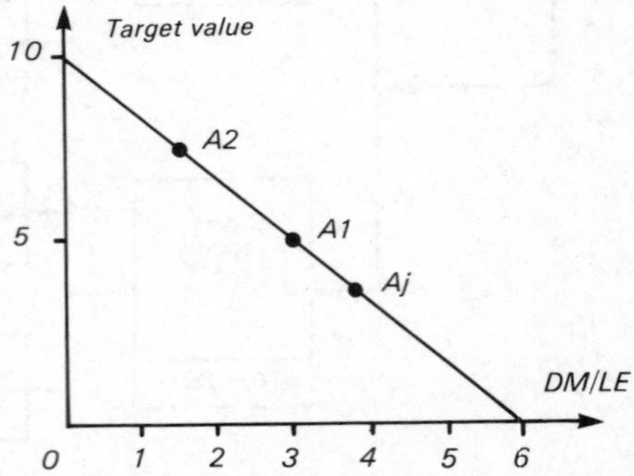 Target value plot: line from (0,10) to (6,0); points A2 ≈ (2, 7), A1 ≈ (3, 5), Aj ≈ (4, 4); x-axis: DM/LE (0,1,2,3,4,5,6); y-axis: 0, 5, 10
⋮ Z_m	. . . .	. . . .

REASONS FOR IMPLEMENTING AN AGVS

In order to fix the weighting scale it is assumed that all systems under comparison are in principle in a position to carry out the transport function in question. Complementary goal definitions in which goals overlap in content (for example, minimum possible operating costs and minimum number of personnel within a goal system) are to be avoided if at all possible or to be accounted for accordingly in the weighting.

Each sub-goal should then be evaluated according to the goal value scale which, for example, can run from 1 to 10. This means that every sub-goal has a certain value depending on the quality of goal fulfilment. Linear functions are usually used for the value functions. The goal value can be determined for each option considered (O_1 to O_j) which is relevant to the decision. With qualitatively defined sub-goals the respective value is estimated. In this connection see Fig. 107 where the procedure for two sub-goals is shown.

After the goal values have been determined for all goals and options they are multiplied by the weighting factors and summed for each option ($O_1 - O_j$).

The sums represent the effectiveness of the final result. This has no designation but represents a rank value. The respective effectiveness values can be incorporated in an industrial management systems approach as an aid to decision making.

With complex transport systems, as with all large projects, it is recommended that the industrial economic investment calculation should be separated from a determination of the effectiveness. In this way a cost-benefit analysis is produced. The advantage of the separated cost-benefit analysis for project options is that only those variables which cannot actually be determined in quantifiable financial terms are evaluated subjectively.

Chapter Nine

Development trends

THE MARKET for AGVSs or driverless transport systems in 1980 can be structured according to the following criteria:

- European market, Federal Republic of Germany market.
- Truck types, for example, pallet trucks, tractors, assembly trucks, and so on.
- Complexity of the AGVS.

Based on the material compiled by the author, but which makes no claim to being totally comprehensive (see appendix), the following conclusions can be drawn:

- Up to the first half of 1980 approximately 360 AGVS installations using a total of approximately 3,900 trucks had been sold on the European market.
- Between 1970 and 1976 the turnover of AGVS sales doubled and between 1977 and 1978 they doubled again; there was no increase in turnover in 1979 and 1980. These data are based on estimated evaluations of the truck costs and the network installation costs per system.
- With respect to turnover, approximately two thirds of the systems sold in Europe are complex systems; this assumes that complex systems are defined as systems with at least 15 trucks or as systems with a central process computer.
- Up to the first half of 1980 inclusively approximately 160 AGVS installations with a total of approximately 920 trucks were sold on the market of the Federal Republic of Germany. From 1970 to 1975 sales remained roughly constant; however, since 1976 sales have doubled. In contrast to the European market 55% of the installations are simple systems and 45% are complex systems; this indicates the high proportion of smaller installations even in medium-sized enterprises.

On the European market special-purpose trucks are quite predominant; pallet trucks and tractors only play a subordinate role. On the other hand on the German market standard pallet trucks and single special-purpose trucks rank roughly equal; tractors are responsible for half the turnover of the two types of trucks mentioned above.

Development trends for AGVSs. The development trends for AGVSs can best be discussed in terms of the three main system areas

☐ Construction of the truck

☐ Traction batteries

☐ Truck control, disposition and data transmission

In terms of truck construction the development of robust and reliable trucks can be considered to be complete. Fully developed units are now in widespread use so that further development will concentrate on minor problems and special load transfer equipment based on customer requirements.

In many industrial sectors, especially in the electrical industry, the weights of internal transport loads have reduced as a result of the introduction, production and assembly of electronic units instead of the previous electrical technology. Consequently there is a certain demand for small trucks with smaller payloads.

Modern heavy traction batteries restrict the application of the AGVS. Although the operating range can be extended by additional charging at intermediate stations one basic relationship remains unchanged: that 25–30% of the weight of the truck is accounted for by the battery.

Despite all efforts the use of batteries with a higher specific power capacity at reasonable costs cannot be foreseen in the immediate future. However, with the help of microprocessor technology and a central process control it has been possible to use AGVS in complex material flow systems; it assumes that there are extensive information transfer facilities throughout the network. Whereas the software for process automation is relatively advanced the transfer of information using inductive guide wires has to be classed as unsatisfactory. The disadvantages of inductive data transmission lie in the effort involved in installation and the associated costs affecting the general flexibility of the entire AGVS installation. The cornerstone of system improvement will therefore be the introduction of radio data transmission systems (Chapter 3).

A completely new perspective on AGVS applications is opened up when they are used in conjunction with industrial robots. In principle applications are possible when, as a result of the system loading, it is more effective to move the robots to several work places (multipoint work) than is the case when the robots are firmly attached to the work place and there is an automatic supply of workpieces.

For such applications the modern types of industrial robot are particularly suitable since their weight more closely matches the needs of the

DEVELOPMENT TRENDS

AGVS. In principle the following applications are possible:

- ☐ Multipoint work in production, for example, servicing of flexible manufacturing systems with relatively long workpiece handling times.
- ☐ Multipoint work in work handling technology, whereby with the help of the AGVS horizontal sections are bridged even during the load handling operation.
- ☐ Commissioning in the field of relatively homogenous product ranges. Here completely new static strategies for presentation and dynamic strategies for the commissioning process have to be developed for representative commissioning systems in view of the robot's mode of movement. It should be noted that the limited access facilities of the robot only permit certain article ranges and that the presentation of articles in the commissioning system must be adapted to these access modes.

Chapter Ten

Summary of AGVS applications

AUTOMATED guided vehicle systems replace and complement conventional industrial trucks over a wide range of applications. The main point of emphasis lies in the field of horizontal transport where AGVS with performances of up to 200 load units an hour compete with conventional stackers, pallet trucks, electric tractors and conveyor systems.

As a result of their technical characteristics, automated load transfer and material flow control by means of intelligent systems on board the trucks of AGVSs are in a position to transport piece parts and load units as individual operations within an extensive network without any personnel being involved. This gives the following advantages to the users of such systems:

- Savings in costs and personnel.
- Automated transport, closing the gap between automated storage systems and manufacturing systems.
- High performance capacity and availability even in the area of high material flow densities, for example at the interface to high-bay stores.
- High transport safety.

The successful application of AGVS is tied to a number of basic prerequisites: those with the highest priority are adequate space; accurate material flow analysis and the consideration of the interfaces at the load transfer points and the material flow control system.

Since 1970 the AGVS has reached a technical level which has enabled it to achieve a rapid penetration of the industrial trucks market so that up to 1980 approximately 160 installations with a total of approximately 920 trucks were sold in the Federal Republic of Germany and approximately 360 installations with a total of approximately 3,900 trucks were sold in Europe.

The most important application areas for DTS are

- Servicing storage and production areas.

- Production-integrated application of DTS trucks as assembly platforms.
- Commissioning, especially in the wholesale trade.
- Servicing special areas such as cold stores, areas with radiation hazards, hospitals and so on.

Furthermore, complex material flow problems with a large number of load transfer stations and interfaces can be solved only with the help of an AGVS using economic automation.

The requirement here is the use of a process control system, preferably with a microprocessor on board the truck as well as an information system for communication between the trucks on the network and the central control system.

Apart from the technical performance capacity of the AGVS its economic viability compared to competing transport systems is an important deciding factor.

An analysis of operating costs with the help of a large number of users of transport systems has showed that for transport operations with material flows of approximately 10 to a maximum of 200 load units an hour and distances of approximately 100 to 3,000 m an AGVS represents an economically viable alternative.

Efforts are being made to make further savings in the costs of internal material flow and continuing improvement in the AGVS, especially through the introduction of radio information transfer, will lead to a widening of the scope of applications.

The existing relatively high flexibility of the AGVS will be additionally improved by radio information transmission because the considerable expense involved in wire-bound transmission is reduced and communication between the truck and the central control system is permanently available.

Completely new applications for the AGVS are possible in combination with industrial robots. In principle the use of a 'driverless' robot is possible for:

- work places in the production area which can be jointly serviced by a robot (multipoint work)
- load handling operations in the storage area, goods-in and dispatch whereby the robot can be employed at different sites. The operating range of the robot's grippers is extended since the truck can also bridge horizontal distances during the handling operation.
- Automated commissioning. New strategies for presenting the articles (static strategies) and for the commissioning process (dynamic strategies) have to be developed here because commissioning technology adapted to humans is considerably improved by the kinematics of the robot. On the other hand the robot has not yet reached the perfection of the human hand.

The further development of the AGVS within the existing range of applications, but especially future applications of the AGVS, close the gap between the automated storage and automated manufacturing and so make further contribution to the development of the automated factory which no longer belongs to the realm of science fiction.

Appendix

AGVS Installations in Europe

THE following statistical data cover the users of AGVS installations insofar as reference lists and other information could be obtained, up to May 1983 inclusively.

The material is arranged according to manufacturer and year of sale. The following characteristics were established for each system:
- User/site
- Branch of industry
- Application
- Transport goods/loading equipment
- Transport weight
- Truck type/truck make
- Network, length, number of switching points, number of stopping points.

Contents

	Page
Installations of Wagner	172
Installations of Jungheinrich	202
Installations of PHB-Babcock	232
Installations of Telelift	234
Installations of Tellus	236
Installations of ACS	244
Installations of BT	249
Installations of CFC	255
Installations of Saxby	256
Installations of Digitron	260
Installations of Barrett	267
Installations of Carrago	271
Installations of FATA	275

YEAR OF SALE	USER/SITE	BRANCH OF INDUSTRY	APPLICATION	TRANSPORT GOODS LOADING EQUIPMENT	Transport weight in tonnes	TRUCK TYPE/ TRUCK MAKE	NETWORK		
							Length in m	Number of switching stops	Number of stopping points
1966	Novibra, 7311 Owen/Tock	Engineering	Servicing production	Small parts in stacking containers	1 t	1 EGN-SO Electric pedestrian-controlled truck	570	6	120
1967	Heller AG 7440 Nürtingen	Engineering	Servicing production	Machine parts on pallets	2 t	7 EGN-SO Electric pedestrian-controlled trucks	760	0	54
1968 1975	Saabergwerke AG, Zentrallager 6600 Saarbrücken	Mining	Distribution and transport in the storage area	Single parts on pallets	2 t	1 EGZ 3000 1 EFZ 3,2 (1975) Electric pedestrian-controlled tractor Electric driver seated tractor	650	8	30
1969 1979	Fordwerke AG 6630 Saarlouis	Car industry	Servicing production	Car parts in skeleton containers	2 t	11 EFZ 9 2 EFZ 12,5 (1979) Electric driver seated tractors	4500	80	44
1970 1978 1979	Siemens AG, Medical equipment department 8520 Erlangen	Metal goods industry	Servicing production	Electrical small parts on pallets	1 t	6 EGU 2000 4 EGU 1200 (1978) 2 EGU 2000 (1979) Electric pedestrian-controlled pallet trucks	2300	5	120

Installations of Wagner, West Germany (Domestic)

APPENDIX

YEAR OF SALE	USER/SITE	BRANCH OF INDUSTRY	APPLICATION	TRANSPORT GOODS LOADING EQUIPMENT	Transport weight in tonnes	TRUCK TYPE/ TRUCK MAKE	NETWORK			
							Length in m	Number of switching stops	Number of stopping points
1971 1977 1978	Heye, 6728 Germersheim	Glass industry	Servicing production	Glass bottles on pallets	1 t	After expansion 1978 10 EFZ 3,2 Electric driver seated tractors	400		
1971	Volkswagenwerk AG 3300 Braunschweig	Car industry	Servicing production	Turned parts in special containers	1 t	4 UFZ 2000 Skid tractors	800	3	according to demand
1972 1975	Deutsche Rhodiaceta AG 7800 Freiburg	Textile industry	Servicing production	Artificial fibres on special trailers	1 t	2 EFZ 3,2 Electric driver seated tractors	920	7	94
1972	Lotter KG, Maschinenfabrik, 7140 Ludwigsburg	Iron wholesale	Servicing production	Iron parts in skeleton container pallets	1,2 t	1 EGU 1200 Electric pedestrian-controlled pallet truck	250	0	2
1972 1977 1978	Heye, 3063 Obernkirchen	Glass industry	Servicing production	Glass bottles on pallets	1 t	After expansion 1979: 19 EFZ 3,2 Electric driver seated tractors	1600	25	60

Installations of Wagner, West Germany (Domestic)

YEAR OF SALE	USER/SITE	BRANCH OF INDUSTRY	APPLICATION	TRANSPORT GOODS LOADING EQUIPMENT	Transport weight in tonnes	TRUCK TYPE/ TRUCK MAKE	NETWORK		
							Length in m	Number of switching stops	Number of stopping points
1972	Melitta-Werke 4950 Minden	Paper processing	Servicing production	Paper on rollers	1 t	1 EGZ 3000 Electric pedestrian-controlled tractor	450	2	9
1972	Rob. Bosch GmbH 3200 Hildesheim	Metal goods industry	Servicing production	Stamped parts in containers	0,5 t	1 EGN 20Q0 Electric pedestrian-controlled truck	100	0	17
1973	Reiche & Co 4937 Lage	Engineering	Servicing production	Mechanical parts on pallets	2 t	2 EGU 2000 R Electric pedestrian-controlled trucks	750	11	20
1973 1979	Herzberger Papierfabrik 3420 Herzberg/Harz	Paper industry	Servicing production	Paper on pallets	1 t	4 EGU 1200 Electric pedestrian-controlled pallet trucks with saddled roller conveyor	200	5	11
1973	Wohlfahrt Elektrogeräte, 7410 Reutlingen	Wholesale	Distribution and transport in the storage areas	Electrical parts in containers	0,5 t	1 EGW Electric pedestrian-controlled truck	200	–	4

Installations of Wagner, West Germany (Domestic)

APPENDIX

| YEAR OF SALE | USER/SITE | BRANCH OF INDUSTRY | APPLICATION | TRANSPORT GOODS LOADING EQUIPMENT | Transport weight in tonnes | TRUCK TYPE/ TRUCK MAKE | NETWORK |||
							Length in m	Number of switching stops	Number of stopping points
1973	Siemens AG, Lager West 8520 Erlangen	Metal goods industry	Distribution and transport in the storage area	Medical equipment on special trailers	3,0 t	Electric driver seated tractors	1800	14	15
1973	Laauser & Co, 7141 Großbottwar	Furniture industry	Servicing of production	Upholstered furniture on roller pallets	0,5 t	1 EGU 1200 Electric pedestrian-controlled pallet truck	250	3	35
1973	ADO Gardinenfabrik 2990 Aschendorf	Textile industry	Passenger transport for monitoring production	Passenger transport	1 t	1 EGW-SO Electric driver seated truck	600	–	–
1974 1979	COOP Central warehouse 2870 Delmenhorst	Chain-store operation	Commissioning	Food on pallets in containers	1 t	2 EGU 1200 Electric pedestrian pallet trucks	300	–	–
1974	Hildesia 3201 Emmerke	Paper processing	Servicing production	Wallpaper on rollers	0,5 t	2 EFZ 3,2 Electric driver seated tractors	700	–	11

Installations of Wagner, West Germany (Domestic)

AUTOMATED GUIDED VEHICLES

YEAR OF SALE	USER/SITE	BRANCH OF INDUSTRY	APPLICATION	TRANSPORT GOODS LOADING EQUIPMENT	Transport weight in tonnes	TRUCK TYPE/ TRUCK MAKE	NETWORK Length in m	Number of switching stops	Number of stopping points
1974	Siemens AG 8000 München	Electronic industry	Servicing production	Electronic modules on pallets	1 t	3 EGU 1200 Electric pallet trucks	220	–	–
1974	AUDI-NSU AG 8070 Ingolstadt	Car industry	Distribution and transport in the storage area	Spare parts on trailers	0,3 t	4 EFZ 12,5 Electric driver seated tractors	2000	30	12
1974	Kugelfischer 8603 Ebern	Ball/roller bearing manufacture	Servicing production	Iron parts on pallets	1 t	2 EGU 2000 Electric pedestrian-controlled pallet trucks	480	2	8
1974 1977	Armstrong-Cork, 4400 Münster	Building materials industry	Servicing production	Insulation materials on pallets	1,5 t	2 EFZ 12,5 Electric driver seated tractors	1650	–	7
1974	Rob. Bosch GmbH 7928 Giengen/Brenz	Electrical industry	Servicing of production	Single parts for household appliances on pallets	1,2 t	9 EGU 1200 Electric pedestrian-controlled pallet trucks	950	13	17

Installations of Wagner, West Germany (Domestic)

APPENDIX 177

YEAR OF SALE	USER/SITE	BRANCH OF INDUSTRY	APPLICATION	TRANSPORT GOODS LOADING EQUIPMENT	Transport weight in tonnes	TRUCK TYPE/ TRUCK MAKE	NETWORK			
							Length in m	Number of switching stops	Number of stopping points
1974	Pfannkuch & Co 7500 Karlsruhe	Food chain-store operation	Commissioning	Food on pallets, loading frames	1 t	5 EGU 1200 Electric pedestrian-controlled pallet trucks	400	—	—
1975	AGA-Radiatorenwerk 2857 Langen	Metal goods industry	Servicing of production	Radiators on pallets	2,5 t	1 EFZ 3,2 Electric driver seated tractor	180	—	—
1975 1980	Fischer 8625 Sonnefeld	Furniture industry	Servicing of production	Upholstered furniture on loading frames	0,4 t	4 EGU 2000 Electric pedestrian-controlled pallet trucks	760	12	60
1975	Gerresheimer 4000 Düsseldorf-Gerresheim	Glass industry	Servicing of production	Glass bottles on pallets	1,2 t	3 EGU-R 2000 Electric pedestrian-controlled pallet trucks	320	9	10
1975 1977	Sachsenwerk 8400 Regensburg	Engineering	Servicing of production	Semi-manufactured goods on special pallets	1,5 t	6 EGN-SO Electric pedestrian-controlled trucks	500	14	36

Installations of Wagner, West Germany (Domestic)

YEAR OF SALE	USER/SITE	BRANCH OF INDUSTRY	APPLICATION	TRANSPORT GOODS LOADING EQUIPMENT	Transport weight in tonnes	TRUCK TYPE/ TRUCK MAKE	NETWORK		
							Length in m	Number of switching stops	Number of stopping points
1976	Zentralkellerei Bad. Winzer 7814 Breisach	Drinks industry	Servicing of production	Wine bottles on pallets	1 t	12 EFZ 3,2 Electric driver seated tractors	2600	56	32
1976 1978	Siemens AG 1000 Berlin	Data technology	Servicing of production	Electric components on pallets, special transport trucks	1 t	4 EFZ 3,2 Electric driver seated tractors	700	7	22
1976	Reichelt 1000 Berlin	Food industry	Commissioning	Food on pallets	0,7 t	30 EGU Electric pedestrian-controlled pallet trucks	1250	8	4
1976	Halberger Hütte 6604 Brebach	Foundry	Servicing of production	Castings on stands	2 t	6 EGN-SO Electric pedestrian-controlled trucks	1200	13	14
1976 1977 1978	Siemens AG 7500 Karlsruhe	Electrical industry	Servicing of production	Electronic parts and computer cabinets on pallets	0,4 t	6 EGU 1250 Electric pedestrian-controlled pallet trucks	750	2	11

Installations of Wagner, West Germany (Domestic)

APPENDIX

YEAR OF SALE	USER/SITE	BRANCH OF INDUSTRY	APPLICATION	TRANSPORT GOODS LOADING EQUIPMENT	Transport weight in tonnes	TRUCK TYPE/ TRUCK MAKE	NETWORK			
							Length in m	Number of switching stops	Number of stopping points
1976 1978	Klaas u. Kock 4432 Gronau	Food wholesale trade	Commissioning	Food on roller containers, pallets	0,5 t 1 t	37 EGHC/5 EGU 1200 Commissioning truck with lift platform/ electric pedestrian-controlled pallet trucks	4200	73	30/ 6
1976	Gerresheimer Glass factory 4000 Düsseldorf-Gerresheim	Glass industry	Servicing of production	Glass bottles on pallets	1,2 t	4 EGU 2000 Electric pedestrian-controlled pallet trucks	185	9	10
1976	BMW AG 8000 München	Car industry	Assembly work	Engines on pallets	0,2 t	35 EG-SO Assembly trucks	500	21	26
1977	Laauser, 7141 Großbottwar	Furniture industry	Distribution and transport in the storage area	Material on rollers	0,05 t	1 EGH 0,05 Electric pedestrian-controlled high-lift truck	400	2	22
1977 1980	Osram, 8900 Augsburg	Electrical industry	Distribution and transport in the storage area	Electrical components on pallets	0,5 t	6 EGU 1200 R Electric pedestrian-controlled pallet trucks	1100	30	28

Installations of Wagner, West Germany (Domestic)

YEAR OF SALE	USER/SITE	BRANCH OF INDUSTRY	APPLICATION	TRANSPORT GOODS LOADING EQUIPMENT	Transport weight in tonnes	TRUCK TYPE/ TRUCK MAKE	NETWORK Length in m	NETWORK Number of switching stops	NETWORK Number of stopping points
1977	Siemens, 8510 Fürth	Electrical industry	Distribution and transport in the storage area	Advertising and information material on pallets	0,8 t	2 EGU 1200 Electric pedestrian-controlled pallet trucks	300	3	12
1978	Baywa Lagerhausges 8000 München	Trading and warehousing operations	Distribution and transport in storage area	Agricultural implements on pallets	1 t	4 EGU 2000 R Electric pedestrian-controlled pallet truck	700	9	22
1978	Westfälische Metall-Industrie 4780 Lippstadt	Metal goods industry	Distribution and transport in the storage area	Production parts on pallets		2 EGU 2000 R Electric pedestrian-controlled pallet trucks	400		
1978	Volkswagen AG 3180 Wolfsburg	Car industry	Assembly work	Bodywork	0,3 t	11 EGY 0,3 Electric skid trucks	320	33	ca.30
1978	König & Flügger 4400 Münster	Chemical industry	Servicing of production	Paint on pallets	1 t	1 EGU 2000 R Electric pedestrian-controlled pallet truck	250		

Installations of Wagner, West Germany (Domestic)

APPENDIX

YEAR OF SALE	USER/SITE	BRANCH OF INDUSTRY	APPLICATION	TRANSPORT GOODS LOADING EQUIPMENT	Transport weight in tonnes	TRUCK TYPE/ TRUCK MAKE	NETWORK Length in m	NETWORK Number of switching stops	NETWORK Number of stopping points
1978	Fraling 4401 Nordwalde	Textile industry	Distribution and transport in the storage area	Textile basic materials on pallets	1 t	1 EGU 2000 R Electric pedestrian-controlled pallet trucks	575		
1978	Heinrich Heine GmbH 7500 Karlsruhe	Gift dispatch	Distribution and transport in the storage area	Gift articles on pallets		7 EGZ 1,0 SO Electric pedestrian-controlled pallet trucks	850	14	18
1978	Wyeth-Pharma GmbH 4400 Münster	Chemical industry	Servicing of production	Pharmaceutical goods on pallets	1 t	1 EGU 2000 Electric pedestrian-controlled pallet trucks	370	–	13
1978	Behr GmbH & Co KG Kühlerfabrik 7130 Mühlacker	Metal goods industry	Servicing of production	Plastic parts, granulates on pallets	1 t	3 EGU 2000 R Electric pedestrian-controlled pallet trucks	580		
1978	Heidelberger Druckmaschinen AG 6908 Wiesloch	Engineering	Servicing of production	Machine parts on commissioning truck		2 EGU 2000 R Electric pedestrian-controlled pallet trucks	300	11	10

Installations of Wagner, West Germany (Domestic)

AUTOMATED GUIDED VEHICLES

YEAR OF SALE	USER/SITE	BRANCH OF INDUSTRY	APPLICATION	TRANSPORT GOODS LOADING EQUIPMENT	Transport weight in tonnes	TRUCK TYPE/ TRUCK MAKE	NETWORK Length in m	Number of switching stops	Number of stopping points
1978	Bühler-Miag GmbH 3300 Braunschweig	Engineering	Servicing of production	Metal parts on pallets		1 EGU 2000 Electric pedestrian-controlled pallet truck	265	4	13
1978	Siemens AG, 7840 Bad Neustadt	Electrical industry	Servicing of production	Motor parts on pallets		2 EGU 2000 Electric pedestrian-controlled pallet trucks	730		
1978	MAN, 8500 Nürnberg	Engineering	Assembly work	Engines on transport stands		4 EGW-ZI/SO 2000 Electric pedestrian-controlled trucks	235		
1978	CO-OP West AG 5020 Frechen	Food chain stores	Commissioning	Luxury articles/ foodstuffs in roller containers		21 EGU 2000 Electric pedestrian-controlled pallet trucks	2900		
1978	Festo Maschinenfabrik 7300 Esslingen	Engineering	Servicing of products	Production parts in transport holders		2 EGHQ-ZI/R 500 Electric pedestrian-controlled stackers	1000		

Installations of Wagner, West Germany (Domestic)

APPENDIX

YEAR OF SALE	USER/SITE	BRANCH OF INDUSTRY	APPLICATION	TRANSPORT GOODS LOADING EQUIPMENT	Transport weight in tonnes	TRUCK TYPE/ TRUCK MAKE	Length in m	NETWORK Number of switching stops	Number of stopping points
1978	Siemens AG 7500 Karlsruhe	Electrical industry	Servicing of production	Electronic parts, computer cabinets on pallets		2 EGU 2000 Electric pedestrian-controlled pallet trucks	400		
1978	Westf. Metallindustrie Zentrallager 4782 Erwitte	Electrical industry	Distribution and transport in storage area	Electrical parts on pallets in containers		9 EGU 2000 R Electric pedestrian-controlled pallet trucks	1700		
1978	Koch, Neff u. Öttinger, 7000 Stuttgart	Publishing distribution warehouse	Distribution and transport in storage area	Books on pallets		2 EGU 2000 R Electric pedestrian-controlled pallet trucks	495		
1979	Raps & Co, 8650 Kulmbach	Spice factory	Servicing of production	Spices on trailers		1 EGZ 2000 Electric pedestrian-controlled tractor	80		
1979 1980	Bosch GmbH, Bühl	Electrical industry	Servicing of production	Electric parts on pallets, in containers		8 EGU 2000 Electric pedestrian-controlled pallet trucks	1100		

Installations of Wagner, West Germany (Domestic)

YEAR OF SALE	USER/SITE	BRANCH OF INDUSTRY	APPLICATION	TRANSPORT GOODS LOADING EQUIPMENT	Transport weight in tonnes	TRUCK TYPE/ TRUCK MAKE	NETWORK			
							Length in m	Number of switching stops	Number of stopping points
1979	H. Hummel GmbH 8625 Sonnefeld	Furniture industry	Distribution and transport in storage area	Upholstered furniture on special stands		3 EGU 1200 Electric pedestrian-controlled pallet trucks	850		
1979	Pfaff GmbH 6750 Kaiserslautern	Engineering	Servicing of production	Metal castings on stands		1 EGZ 2000 Electric pedestrian-controlled tractor	150		
1979	MAN AG 8900 Augsburg	Engineering	Servicing of production	Metal parts in transport containers		3 EGN 2000 Electric pedestrian-controlled trucks	320		
1979	Schubert & Salzer, AG 8070 Ingolstadt	Engineering	Servicing of production	Metal castings on stands		1 EGU 2000 R Electric pedestrian-controlled pallet truck	120		
1979	Brandstetter GmbH & Co KG, 8502 Zirndorf	Toy industry	Servicing of production	Semi-manufactured and finished products on pallets		1 EGU 1200 Electric pedestrian-controlled pallet truck	150		

Installations of Wagner, West Germany (Domestic)

APPENDIX

YEAR OF SALE	USER/SITE	BRANCH OF INDUSTRY	APPLICATION	TRANSPORT GOODS LOADING EQUIPMENT	Transport weight in tonnes	TRUCK TYPE/ TRUCK MAKE	NETWORK		
							Length in m	Number of switching stops	Number of stopping points
1979	Ford-Werke AG 5000 Köln	Car industry	Distribution and transport in storage area	Car parts on special trailers		5 EFZ 12,5 Electric driver seated tractors	1200		
1979	Allianz-Versicherungs-AG, 7000 Stuttgart	Insurance management	Document transport	Documents on special stands		3 EGN 01-SO Electric document transporters	1750		
1979	Daimler Benz AG 7032 Sindelfingen	Car industry	Distribution and transport in the storage area	Spare parts on special trailers		12 EFZ 12,5 Electric driver seat tractors	3500		
1979	Volkswagenwerk AG 3320 Salzgitter	Car industry	Servicing of production	Turned parts on trailers		5 EGZ 1500 Electric pedestrian-controlled tractors	950		
1979 1980	Bünting & Co 2950 Leer	Food industry	Commissioning	Foodstuffs and luxury articles on roller containers		15 EGU 2000 Electric pedestrian-controlled pallet trucks	1700		

Installations of Wagner, West Germany (Domestic)

AUTOMATED GUIDED VEHICLES

YEAR OF SALE	USER/SITE	BRANCH OF INDUSTRY	APPLICATION	TRANSPORT GOODS LOADING EQUIPMENT	Transport weight in tonnes	TRUCK TYPE/ TRUCK MAKE	NETWORK		
							Length in m	Number of switching stops	Number of stopping points
1979	Sonopress/Bertelsmann AG 4830 Gütersloh	Record production	Servicing of production	Record packaging material on pallets		2 EGU 2000 R Electric pedestrian-controlled pallet trucks	350		
1980	Trumpf GmbH, Maschinenfabrik, 7257 Ditzingen	Engineering	Servicing of production	Machine parts on pallets in containers		1 EGU-R 2000 Electric pedestrian-controlled pallet truck	500		
1980	Trumpf GmbH, Maschinenfabrik, 7257 Ditzingen	Engineering	Servicing of production	Metal cassettes	3 t	1 EG-SO 3000 Electric pedestrian-controlled skid	80		
1980	Meistermarkenwerke Delmenhorst	Food industry	Distribution and transport in storage area	Cardboard boxes with margarine on pallets		4 EGU 2000 Electric pedestrian-controlled pallet trucks	400		
1980	Volkswagenwerk AG 3180 Wolfsburg	Car industry	Servicing of production	Turned parts on trailers		4 EGZ 1500 Electric pedestrian-controlled trucks	800		

Installations of Wagner, West Germany (Domestic)

APPENDIX

YEAR OF SALE	USER/SITE	BRANCH OF INDUSTRY	APPLICATION	TRANSPORT GOODS LOADING EQUIPMENT	Transport weight in tonnes	TRUCK TYPE/ TRUCK MAKE	NETWORK		
							Length in m	Number of switching stops	Number of stopping points
1980	Kalle AG 6200 Wiesbaden	Foil production	Servicing of production	Foil rolls on roller stands	2,5 t	2 EG-SO Electric pedestrian-controlled skids	230		
1980	Kalle AG 6680 Neunkirchen	Foil production	Servicing of production	Foil rolls and empty cores on roller stands	2,5 t	2 EG-SO Electric pedestrian-controlled skids	600		
1980	Barmag 5630 Remscheid	Engineering	Servicing of production	Machine parts in wooden containers on pallets		1 EGU 2000 R Electric pedestrian-controlled pallet truck	220		
1980	Zentralkellerei Bad. Winzer 7814 Breisach	Drinks industry	Passenger transport	Visitors on trailers		2 EFZ 12 Electric driver seated tractors	2400		
1980	Bosch-Siemens Hausgeräte GmbH 8510 Fürth	Central stores for spare parts	Distribution and transport in the storage area	Spare parts on pallets/skeleton containers		2 EGU 2000 R Electric pedestrian-controlled pallet trucks	860		

Installations of Wagner, West Germany (Domestic)

AUTOMATED GUIDED VEHICLES

YEAR OF SALE	USER/SITE	BRANCH OF INDUSTRY	APPLICATION	TRANSPORT GOODS LOADING EQUIPMENT	Transport weight in tonnes	TRUCK TYPE/ TRUCK MAKE	NETWORK		
							Length in m	Number of switching stops	Number of stopping points
1980	Robert Bosch GmbH 7000 Stuttgart-Feuerbach	Engineering	Servicing of production	Fuel pumps on special trailers		1 EGZ 4000 Electric pedestrian-controlled pallet truck	500		
1980	Mann & Hummel 8311 Marklkofen	Filter works	Servicing of production	Machined parts in skeleton containers		2 EGU 2000 R Electric pedestrian-controlled pallet trucks	1200		
1980	Edeka Handels GmbH Balingen	Food	Distribution and transport in storage area	Food on pallets		4 EGU 2000 R Electric pedestrian-controlled pallet trucks	700		
1980	Paul Hartmann AG 7920 Heidenheim	Textile industry	Servicing of production	Cellulose rolls cardboxes on pallets		3 EGU 1250 Electric pedestrian-controlled pallet trucks	650		
1980	Siemens AG 8740 Bad Neustadt	Electric industry	Servicing of production	Pallets and skeleton containers		2 EGU 2000 Electric pedestrian-controlled pallet trucks	620		

Installations of Wagner, West Germany (Domestic)

APPENDIX

| YEAR OF SALE | USER/SITE | BRANCH OF INDUSTRY | APPLICATION | TRANSPORT GOODS LOADING EQUIPMENT | Transport weight in tonnes | TRUCK TYPE/ TRUCK MAKE | NETWORK |||
							Length in m	Number of switching stops	Number of switching stopping points
1981	Friedrich-Deckel AG 8000 München	Engineering	Servicing of production	Composite wood		1 EGN-SO-ZI	80		
1982	Bosch 7410 Reutlingen	Automobile industry	Distribution and transport in storage areas	Reflectors on pallets	2	1 EGL-R-ZI	100		
1982	Planet-Möbel 2910 Westerstede	Furniture industry	Distribution and transport in storage areas	Furniture on special pallets	1,6	3 EGU-R-ZI	150		
1981	Andreas Stihl 7050 Waiblingen	Engineering	Servicing of production	Machine parts on pallets and boxes	1	2 EGZ-ZI-R	450		
1982	Volkswagen AG 2970 Emden	Automobile industry	Servicing of production	Motor cars	1,2	24 EGU-SO-ZI	5000		

Installations of Wagner, West Germany (Domestic)

YEAR OF SALE	USER/SITE	BRANCH OF INDUSTRY	APPLICATION	TRANSPORT GOODS LOADING EQUIPMENT	Transport weight in tonnes	TRUCK TYPE/ TRUCK MAKE	NETWORK		
							Length in m	Number of switching stops	Number of stopping points
1983	BJB Böckelmann 5770 Arnsberg	Engineering	Servicing of production	Plastic parts and metal items on pallets	1	1 EGSM-X-ZI	300		
1983	Volkswagen AG 2970 Emden	Automobile industry	Battery loading station	Batteries on special pallets	0,6	1 EGN-R-ZI	100		
1983	Wiegand & Söhne Steinbach			Glass bottles on pallets		3 EGW-R-ZI	600		

Installations of Wagner, West Germany (Domestic)

APPENDIX

| YEAR OF SALE | USER/SITE | BRANCH OF INDUSTRY | APPLICATION | TRANSPORT GOODS LOADING EQUIPMENT | Transport weight in tonnes | TRUCK TYPE/ TRUCK MAKE | NETWORK |||
							Length in m	Number of switching stops	Number of stopping points
1969	Swedform, Skillingargd/Sweden	Furniture industry	Distribution and transport in the storage area	Furniture on pallet trucks		1 EGZ 2000 Electric pedestrian-controlled tractor	200		
1971	Bechler S.A. Maschinenfabrik, Moutier/Switzerland	Engineering	Servicing of production	Workpieces on pallets		1 EGU 1200 Electric pedestrian-controlled pallet truck	400		
1971	Installation in the GDR	Electrical industry	Servicing of production	Magnetic tapes on trailers		4 EFZ 3 Electric driver seated tractors	650		
1971	Janssen, Beerse/Belgium	Chemical industry	Distribution and transport in storage area	Medicaments on pallets		2 EGU 1200 Electric pedestrian-controlled pallet trucks	350		
1972	Manufacturer of chemical products Luxembourg	Chemical industry	Servicing of production	Film material on trailers		6 EFZ 3000 Electric driver seated tractors	800		

Installations of Wagner, West Germany (Elsewhere)

YEAR OF SALE	USER/SITE	BRANCH OF INDUSTRY	APPLICATION	TRANSPORT GOODS LOADING EQUIPMENT	Transport weight in tonnes	TRUCK TYPE/ TRUCK MAKE	NETWORK Length in m	Number of switching stops	Number of stopping points
1972	Rieter AG, Maschinenfabrik, Winterthur/Switzerland	Engineering	Servicing of production	Machined parts on pallets in containers		12 EGU 2000 Electric pedestrian-controlled pallet trucks	1400		
1972	Rieter AG, Maschinenfabrik, Winterthur/Switzerland	Engineering	Servicing of production	Workpieces in special roller containers		1 EGU 1200 Electric pedestrian-controlled pallet truck	300		
1973	Manuli, Castelforte/Italy	Textile industry	Servicing of production	Foils on pallets		2 EFZ 3,2 Electric driver seated tractors	1000		
1973	Installation in the GDR	Engineering	Servicing of production	Springs on pallets		2 EGU 1200/1 EFZ 3,2 Electric pedestrian-controlled pallet trucks Electric driver seated tractor	800		
1973	Installation in the GDR	Engineering	Servicing of production	Mechanical parts on trailer		5 EFZ 3,2/7 EFZ 12,5 Electric driver seated tractors	250		

Installations of Wagner, West Germany (Elsewhere)

APPENDIX

| YEAR OF SALE | USER/SITE | BRANCH OF INDUSTRY | APPLICATION | TRANSPORT GOODS LOADING EQUIPMENT | Transport weight in tonnes | TRUCK TYPE/ TRUCK MAKE | NETWORK |||
							Length in m	Number of switching stops	Number of stopping points
1974	van Nelle, Rotterdam/Holland	Luxury article industry	Distribution and transport in the storage area	Tobacco/tea on pallets		3 EGU-1200 R Electric pedestrian-controlled pallet trucks	750		
1974	Philips, Barcelona/Spain	Electric industry	Servicing of production	Electrical articles on pallets		4 EGU 1600 Electric pedestrian-controlled pallet trucks	735		
1974	Enka-Werke, Emmen/Holland	Textile industry	Servicing of production	Fibres in cans		1 EG-SO-R Transport truck and conveyor	220		
1974	Philips N.V., Leuven/Holland	Electrical industry		Radio sets on pallets		3 EGU 1200 Electric pedestrian-controlled pallet trucks	320		
1975	SIP, Turin/Italy	Telecommunications	Servicing of production	Electrical parts on pallets		2 EGU 2000-R Electric pedestrian-controlled pallet trucks	650		

Installations of Wagner, West Germany (Elsewhere)

YEAR OF SALE	USER/SITE	BRANCH OF INDUSTRY	APPLICATION	TRANSPORT GOODS LOADING EQUIPMENT	Transport weight in tonnes	TRUCK TYPE/ TRUCK MAKE	NETWORK Length in m	NETWORK Number of switching stops	NETWORK Number of stopping points
1975	Chr. Dior Orleans/France	Cosmetics industry	Distribution and transport in the storage area	Cosmetics articles on pallets		4 EFZ 3,2 Electric driver seated tractors	660		
1975	Bruynzeel B.V. Zaandam/Holland	Timber industry		Doors on pallets		4 EFZ 3,2 Electric driver seated tractors	660		
1975	Saab-Scania Södertälje/Sweden	Car industry	Servicing of production	Lorry engines on loading stands		2 EGW 2000 Electric pedestrian-controlled trucks	200		
1975	ALSO, Naples/Italy	Food and luxury article industry	Distribution and transport in the storage area	Ice cream on trailers		4 EFZ 3,2 Electric driver seated tractors	870		
1975	Behr (Radiator factory), Roufach/France	Metal goods industry	Servicing of production	Punched parts and castings on pallets in containers		7 EGU 2000 Electric pedestrian-controlled pallet trucks	480		

Installations of Wagner, West Germany (Elsewhere)

APPENDIX

YEAR OF SALE	USER/SITE	BRANCH OF INDUSTRY	APPLICATION	TRANSPORT GOODS LOADING EQUIPMENT	Transport weight in tonnes	TRUCK TYPE/ TRUCK MAKE	NETWORK Length in m	Number of switching stops	Number of stopping points
1976	Colruyth N.V. Halle/Belgium	Food wholesale trade	Commissioning	Food on commissioning trucks		30 EGU 2000 Electric pedestrian-controlled pallet trucks	1250		
1976	General Motors Antwerp/Belgium	Car industry	Servicing of production	Assembly parts on pallets		3 EGZ 400 Electric pedestrian-controlled tractors	1400		
1976	Van der Werff B.V. Leek/Holland	Food wholesale trade	Commissioning	Food on roller containers		14 EGU 2000 Electric pedestrian-controlled pallet trucks	440		
1976	Montefibre Vercelli/Italy	Chemical industry	Servicing of production	Chemical products on pallets		2 EGU 2000 R Electric pedestrian-controlled pallet trucks	480		
1976	Montedison Massa Carrara/Italy	Chemical industry	Servicing of production	Chemical products on pallets		4 EGN with built-on roller conveyor	800		

Installations of Wagner, West Germany (Elsewhere)

| YEAR OF SALE | USER/SITE | BRANCH OF INDUSTRY | APPLICATION | TRANSPORT GOODS LOADING EQUIPMENT | Transport weight in tonnes | TRUCK TYPE/ TRUCK MAKE | NETWORK |||
							Length in m	Number of switching stops	Number of stopping points
1977	Bloemensveiling Ver. Norden, Elde/Holland	Flower auctioning	Distribution and transport	Cut flowers, pot plants on tier trailers		7 EGZ 2000 ZI Electric pedestrian-controlled tractor with driver platform	260		
1978	Snikkers, Strijen/Holland	Sanitary wholesale trade	Distribution and transport in storage area	Sanitary articles on pallets		1 EGU 2000 R Electrical pedestrian-controlled pallet truck	120		
1978	Philips N.V. Eindhoven/Holland	Electrical industry		Television sets		8 EG-ZI/SO Electric tractor	820		
1974	Source Perrier Vergèse/France	Drinks industry	Distribution and transport in storage area	Bottles with mineral water		50 EG-ZI-SO 50 Electric tractors	4500		
1978	Hotel Pichler Prissiano/South Tyrol	Hotel and catering industry	Transport of meals	Meals/drinks on trays		7 EGN-SO Electric meal transport trucks	170		

Installations of Wagner, West Germany (Elsewhere)

APPENDIX

| YEAR OF SALE | USER/SITE | BRANCH OF INDUSTRY | APPLICATION | TRANSPORT GOODS LOADING EQUIPMENT | Transport weight in tonnes | TRUCK TYPE/ TRUCK MAKE | NETWORK |||
							Length in m	Number of switching stops	Number of stopping points
1978	Euroffice Pomezia/Italy	Paper industry	Servicing of production	Paper on pallets		3 EGN-R Electric skid tractors	600		
1978	Cartiere Burgo Corsico/Italy	Paper industry	Distribution and transport in the storage area	Paper rolls on trailers		1 EFZ 15-ZI/R Electric drive seated tractors	180		
1978	Moplefan Terni/Italy	Chemical industry	Distribution and transport in the storage area	Polypropylene on pallets		2 EGU 1200 1 EGV 1200-ZI/R SO Electric pedestrian-controlled pallet trucks Electric pedestrian-controlled high-lift truck	1000		
1978	Minerva Fittings Opava/Czechoslovakia	Metal goods industry	Servicing of production	Production parts on pallets		5 EG-ZI 0,4 Electric transport trucks	750		
1978	CZM-Motorcycles Strokonice/Czechoslovakia	Motorcycle industry	Servicing of production	Machine parts on pallets		5 EG-ZI 0,4 Electric pedestrian-controlled transport vehicles	750		

Installations of Wagner, West Germany (Elsewhere)

YEAR OF SALE	USER/SITE	BRANCH OF INDUSTRY	APPLICATION	TRANSPORT GOODS LOADING EQUIPMENT	Transport weight in tonnes	TRUCK TYPE/ TRUCK MAKE	NETWORK Length in m	Number of switching stops	Number of stopping points
1978	Sigma, Zavadka Zadvadka/ Czechoslovakia	Metal goods industry	Servicing of production	Production parts on pallets		5 EG-ZI 0,4 Electric transport trucks	50		
1979	Gualchierani S.p.A. Vercelli/Italy			Special transport truck		5 EG-SO 2000 Electric pedestrian-controlled pallet trucks	1200		
1979	Flower auctioning Berkel/Holland	Flower auctioning	Distribution and transport	Cut flowers, pot plants on tier trailers		5 EGZ 2000-SO Electric pedestrian-controlled tractors	250		
1979	Anic, Pisticci/Italy	Chemical industry	Servicing of production	Special containers		4 EGW-SO 1500 Electric pedestrian-controlled truck	450		
1979	ZTS (Utar) Bratislava/ Czechoslovakia	Scientific institute	Linking of NC machines	Clamping pallets		1 EG-SO 0,8 Electric pedestrian-controlled transport vehicle	180		

Installations of Wagner, West Germany (Elsewhere)

APPENDIX

YEAR OF SALE	USER/SITE	BRANCH OF INDUSTRY	APPLICATION	TRANSPORT GOODS LOADING EQUIPMENT	Transport weight in tonnes	TRUCK TYPE/ TRUCK MAKE	NETWORK Length in m	NETWORK Number of switching stops	NETWORK Number of stopping points
1980	Tatra Koprivnice/ Czechoslovakia	Car industry		Machine parts on pallets and transport stands		8 EGU 1200 R Electric pedestrian-controlled pallet trucks	1800		
1980	VEB Biesenthal/GDR	Furniture industry	Distribution and transport in the storage area	Foil		1 EGN-R Electric pedestrian-controlled truck	60		
1980	VEB Orbitaplast Weissandt-Golzau/GDR	Plastics industry	Servicing of production	Plastic foil on roller stands		5 EGU-R 2000 Electric pedestrian-controlled trucks	1100		
1980	Anic/Sardinia	Chemical industry		Special container		10 EGW-SO 1500 Electric pedestrian-controlled trucks	1200		
1980	Gullfiber Söraker/Sweden	Building material industry	Distribution and transport in the storage area	Insulation materials on trailers		11 EFZ 1200 Electric driver seat tractors	3000		

Installations of Wagner, West Germany (Elsewhere)

YEAR OF SALE	USER/SITE	BRANCH OF INDUSTRY	APPLICATION	TRANSPORT GOODS LOADING EQUIPMENT	Transport weight in tonnes	TRUCK TYPE/ TRUCK MAKE	Length in m	NETWORK	
								Number of switching stops	Number of stopping points
1980	Mazuchelli Varese/Italy	Plastics industry	Distribution and transport in storage areas	Plastic rolls on pallets		5 EGU 2000 R	350		
1980 1981	SIAI Marchetti Varese/Italy	Aerospace industry	Servicing of production	Aircraft parts on trailers		1 EGZ 4000 SO 1 EGZ 2000	380		
1981	Rhone-Poulene-Specia St. Genis-Laval	Pharmaceuticals	Distribution and transport in storage areas	Pharmaceutical products		5 EGU 2000 R	820		
1981	L'oreal-Saipo Turin/Italy	Cosmetics industry		Cosmetics on pallets		5 EG-50	250		
1981	Installation in the USSR	Textile Industry	Servicing of production	Cloth on rolls		14 EGN 2000 R	6875		

Installations of Wagner, West Germany (Elsewhere)

APPENDIX

YEAR OF SALE	USER/SITE	BRANCH OF INDUSTRY	APPLICATION	TRANSPORT GOODS LOADING EQUIPMENT	Transport weight in tonnes	TRUCK TYPE/ TRUCK MAKE	NETWORK		
							Length in m	Number of switching stops	Number of stopping points
1981	ZTS Hetvikov/CSSR	Engineering	Servicing of production	Machined parts on special pallets		2 EG-SO	550		
1981	Thomassen Drijver Verblitten Doesburg/Holland	Steel industry		Cast iron on pallets		1 EGU 2000 R	670		

Installations of Wagner, West Germany (Elsewhere)

YEAR OF SALE	USER/SITE	BRANCH OF INDUSTRY	APPLICATION	TRANSPORT GOODS LOADING EQUIPMENT	Transport weight in tonnes	TRUCK TYPE/ TRUCK MAKE	NETWORK Length in m	NETWORK Number of switching stops	NETWORK Number of stopping points
1966	Gebr. Claas, 4834 Hardsewinkel	Truck construction	Servicing of production	Agricultural machinery parts in skeleton container	0,1 t	1 EDZ-HF 40 1 EDZ-HF 80 1 EDZ-I 50 (1980) Electric pole tractors	700	6	40
1966	Croon & Lucke Maschinenfabrik GmbH, 7947 Mengen	Engineering	Servicing of production	Long articles	0,5 t	1 EDZ-HF Electric pole tractor	390	–	12
1968	Lorenz AG, 7505 Ettlingen	Engineering	Servicing of production	Machine parts on pallets	0,5 t	2 EDZ-HF 40 Electric pole tractors	450	4	53
1969 1973	Storebest GmbH, 24 Lübeck 1	Furniture industry	Servicing of production	Furniture on pallets	1 t	1 EDZ-HF 40 3 EDZ-HF 80 Electric pole tractors	1400	19	24
1969	Meyerhoff, 2860 Osterholz-Scharmbeck	Furniture industry	Servicing of production	Furniture on pallets	0,3 t	1 EDZ-HF 40 Electric pole tractor	380	1	12

Installations of Jungheinrich, West Germany (Domestic)

APPENDIX

| YEAR OF SALE | USER/SITE | BRANCH OF INDUSTRY | APPLICATION | TRANSPORT GOODS LOADING EQUIPMENT | Transport weight in tonnes | TRUCK TYPE/ TRUCK MAKE | NETWORK |||
							Length in m	Number of switching stops	Number of stopping points
1969	Blendax Werke GmbH 65 Mainz 1	Cosmetics industry	Servicing of production	Packaging material on pallets	0,5 t	1 EDZ-HF 40 Electric pole tractor	380	1	10
1969	Westf. Spinnerei und Kunststoffwerk GmbH 4950 Minden	Textile industry	Servicing of production	Sisal (fibre) in containers	0,5 t	1 EDZ-HF 20 Electric pole tractor	190	–	5
1969 1979	Tehalit Kunststoffwerk GmbH, 6751 Heltersberg	Chemical industry	Servicing of production	Plastic sections on built-up pallets	1,2 t	2 EDZ-HF 80 1 EGZ-I-150 (1979) Electric pole tractors	800 600 (1979)		
1970	Volkswagenwerk, 3321 Salzgitter-Beddingen	Car industry	Servicing of production	Engine parts, sheet metal in box pallets	2 t	6 EFZ-HF 120 Electric driver seated tractors	1800	25	29
1970	Gustav Hensel KG 5940 Lennestadt-Altenhünden	Electrical industry	Servicing of production	Electrical parts on pallets	0,8 t	2 EDZ-HF 20 Electric pole tractors	220	1	17

Installations of Jungheinrich, West Germany (Domestic)

YEAR OF SALE	USER/SITE	BRANCH OF INDUSTRY	APPLICATION	TRANSPORT GOODS LOADING EQUIPMENT	Transport weight in tonnes	TRUCK TYPE/ TRUCK MAKE	NETWORK		
							Length in m	Number of switching stops	Number of stopping points
1971	G. & J. Jäger GmbH 56 Wuppertal-Elberfeld	Engineering	Servicing of production	Rolling bearing parts in containers	0,3 t	2 EDZ-HF 40 Electric pole tractors	500	3	5
1971	Maschinenfabrik Hans Lenze KG 4923 Bösingfeld	Engineering	Servicing of production	Machine parts on pallets	1 t	1 EDZ-HF 80 Electric pole tractor	400	–	8
1971	Trumpf & Co 7257 Ditzingen	Engineering	Servicing of production	Machine parts on pallets	1 t	1 EDZ-HF 20 Electric pole tractor	390	3	47
1971	P.I.V. Antrieb Werner Reimers KG 638 Bad Homburg	Engineering	Servicing of production	Gear parts on pallets	0,6 t	1 EDZ-HF 20 Electric pole tractor	550	3	10
1971	Val. Mehler AG 64 Fulda	Weaving mill	Servicing of production	Cord bales	1,5 t	1 EDZ-HF 20 Electric pole tractor	250	–	8

Installations of Jungheinrich, West Germany (Domestic)

APPENDIX

YEAR OF SALE	USER/SITE	BRANCH OF INDUSTRY	APPLICATION	TRANSPORT GOODS LOADING EQUIPMENT	Transport weight in tonnes	TRUCK TYPE/ TRUCK MAKE	Length in m	NETWORK Number of switching stops	Number of stopping points
1972 1979	Deutsche Lufthansa AG 2 Hamburg 63	Aviation	Distribution and transport in storage area	Aircraft parts on pallets	1 t	4 EFZ-120 1 EFZ-120 (1979) Electric driver seated tractors	1500 600 (1979)	3	22
1972	Veba-Glas AG 43 Essen 12	Glass industry	Servicing of production	Glass bottles on pallets	1,5 t	9 EFZ-HF 120 Electric driver seated tractors	3000	16	14
1972	Argo GmbH 7521 Menzingen	Engineering	Servicing of production	Machine parts on pallets	0,5 t	1 EDZ-HF 20 Electric pole tractor	250	3	33
1972	Präwema GmbH & Co 344 Eschwege	Engineering	Servicing of production	Machine parts on pallets	1 t	2 EDZ-HF 20 Electric pole tractor	300	2	42
1973 1980	Adolf Illig 71 Heilbronn-Sontheim	Engineering	Servicing of production	Machine parts on pallets	1 t	1 EDZ-HF 20 Electric pole tractor	320 140 (1980)	–	12

Installations of Jungheinrich, West Germany (Domestic)

YEAR OF SALE	USER/SITE	BRANCH OF INDUSTRY	APPLICATION	TRANSPORT GOODS LOADING EQUIPMENT	Transport weight in tonnes	TRUCK TYPE/ TRUCK MAKE	NETWORK Length in m	Number of switching stops	Number of stopping points
1973	Kleber-Colombes AG 667 St. Ingbert	Chemical industry	Servicing of production	Tyres	1,5 t	2 EDZ-HF 120 Electric pole tractors	850	4	4
1973	Brüggershemke & Reinkemeier KG 4830 Gütersloh	Paper industry	Servicing of production	Office articles on pallets	0,5 t	1 EDZ-HF 20 Electric pole tractor	280	1	3
1973	Kaufring 4 Düsseldorf 30	Department store	Distribution and transport in the storage area	Department store articles on pallets	0,5 t	10 EFZ-HF 40 3 EFZ-I 50 (1978) Electric driver seated tractors	2200 1100 (1978)	20	47
1973 1975	CO-OP Dortmund 46 Dortmund-Brackel	Food warehouse	Commissioning	Food on roller containers	0,5 t	5 EJE-HF 20 22 EJA-HF 4+1 (1975) 4 EJA-HF 4+1 (1978) Commissioning pallet trucks	3000	51	45
1973 1978	IBM Deutschland GmbH 1 Berlin 48	Electrical industries	Servicing of production	Copiers, typewriters on pallets	1 t	2 EDZ-HF 40 1 EGZ-I 50 (1978) Electric pole tractors	2000	15	15

Installations of Jungheinrich, West Germany (Domestic)

APPENDIX

YEAR OF SALE	USER/SITE	BRANCH OF INDUSTRY	APPLICATION	TRANSPORT GOODS LOADING EQUIPMENT	Transport weight in tonnes	TRUCK TYPE/ TRUCK MAKE	NETWORK		
							Length in m	Number of switching stops	Number of stopping points
1973	Siemens AG 85 Nürnberg	Electrical industry	Servicing of production	Control cabinets on special stands	1 t	1 EJE-HF 20 Electric pole truck	100	–	7
1973	Siemens AG 85 Nürnberg	Electrical industry	Servicing of production	Semi-manufactured goods on special stands	0,4 t	1 EJE-HF 20 Electric pole pallet truck	100	–	8
1973	Siemens AG 581 Witten	Electrical industry	Servicing of production	Control cabinets on trucks	1 t	5 EDZ-HF 20 Electric pole trucks	1230	7	15
1974	Siemens AG 6140 Bensheim	Electrical industry	Servicing of production	Electrical parts on pallets	0,3 t	1 EJE-HF 20 Electric pole pallet truck	470	–	11
1974	Neue Argus GmbH 7507 Ettlingen	Engineering	Servicing of production	Machine parts on pallets	2,5 t	2 EDZ-HF 20 Electric pole tractors	950	10	54

Installations of Jungheinrich, West Germany (Domestic)

AUTOMATED GUIDED VEHICLES

YEAR OF SALE	USER/SITE	BRANCH OF INDUSTRY	APPLICATION	TRANSPORT GOODS LOADING EQUIPMENT	Transport weight in tonnes	TRUCK TYPE/ TRUCK MAKE	NETWORK		
							Length in m	Number of switching stops	Number of stopping points
1974	Berkel 4100 Duisburg	Engineering	Servicing of production	Scales on pallets	0,5 t	1 EDZ-HF 20 Electric pole tractor	260	–	7
1974	Kaufhof AG 505 Porz	Department store	Distribution and transport in the storage area	Department store articles	0,4 t	5 EDZ-HF 20 Electric pole tractors	1700	15	17
1974	Federal army depot		Tunnel application	Pallets	0,3 t	3 EDZ-HF 20 7 EDZ-HF 80 Electric pole tractor	5500	21	43
1975	Möbel-Kraft 236 Bad Segeberg	Furniture industry	Servicing of production	Furniture	0,5 t	4 EFZ-HF 40 Electric pole tractors	680	18	12
1975	Karl Kässbohrer GmbH 79 Ulm	Truck construction	Servicing of production	Machine parts on pallets	0,5 t	1 EGU-I 30 Electric pole pallet truck	830	11	70

Installations of Jungheinrich, West Germany (Domestic)

APPENDIX

YEAR OF SALE	USER/SITE	BRANCH OF INDUSTRY	APPLICATION	TRANSPORT GOODS LOADING EQUIPMENT	Transport weight in tonnes	TRUCK TYPE/ TRUCK MAKE	Length in m	NETWORK Number of switching stops	Number of stopping points
1975	Pfaff Industrie- maschinen, 6750 Kaiserslautern	Engineering	Servicing of produc- tion	Machine parts in bins	0,5 t	1 EGZ-I 20 Electric tractor	180	1	13
1976	Süddeutsche Metall- werke GmbH 6909 Walldorf	Engineering	Servicing of produc- tion	Semi-manu- factured goods on pallets	0,8 t	1 EFZ-I 100 Electric driver seat tractor	500	–	6
1976	Trumpf & Co., Werk II 7481 Hettingen	Engineering	Servicing of produc- tion	Machines	7,5 t	1 EGZ-I 150 Electric tractor	400	2	8
1976	Volkswagenwerk 3321 Salzgitter- Beddingen	Car industry	Assembly work	Engines on special stands	0,25t	30 EGP-I-Sonder Special assembly trucks	300	2	8
1976	Adolf Klinge & Co 9 München	Chemical industry	Servicing of produc- tion	Chemical products in skeleton con- tainers	0,7 t	1 EGZ-I 50 Electric tractor	400	2	4

Installations of Jungheinrich, West Germany (Domestic)

YEAR OF SALE	USER/SITE	BRANCH OF INDUSTRY	APPLICATION	TRANSPORT GOODS LOADING EQUIPMENT	Transport weight in tonnes	TRUCK TYPE/ TRUCK MAKE	NETWORK Length in m	NETWORK Number of switching stops	NETWORK Number of stopping points
1976	E. Merck 61 Darmstadt	Chemical industry	Servicing of production	Chemicals in aluminium and plastic cases	1,5 t	1 EGZ-I 50 Electric tractor	600	2	2
1977	Kalle 6202 Wiesbaden-Biebrich	Chemical industry	Servicing of production	Foils on rollers	1,5 t	1 EFZ-I 20 Electric driver seated tractor	180	–	4
1977	Kaufhof AG 8 München	Department store	Distribution and transport in storage area	Department store on roller containers	0,6 t	10 EGZ-I 50 Electric tractors	2000	18	21
1977	H. Meyer & Co KG Berlin-Reinickendorf	Food industry	Commissioning	Food on roller containers	0,5 t	15 EGZ-IA Electric tractor	1400	18	13
1977	BMW 8312 Dingolfing	Spare parts store	Distribution and transport in the storage area	Spare parts on pallets	1 t	28 EGU-I 15 3 EGU-I 15 (1980) Electric pole pallet trucks	2500 1000 (1980)	57	50

Installations of Jungheinrich, West Germany (Domestic)

APPENDIX

YEAR OF SALE	USER/SITE	BRANCH OF INDUSTRY	APPLICATION	TRANSPORT GOODS LOADING EQUIPMENT	Transport weight in tonnes	TRUCK TYPE/ TRUCK MAKE	NETWORK			
							Length in m	Number of switching stops	Number of stopping points
1977	Schroff KG Industriegebiet 7541 Straubenhardt 1	Electrical industry	Servicing of production	Semi-manufactured goods on pallets	1,5 t	2 EFZ-I 50 Electric driver seat tractors	500	3	3
1977	Mann & Hummel GmbH 7140 Ludwigsburg	Engineering	Servicing of production	Machine parts on pallets	1,5 t	2 EGZ-I 100 Electric tractors	500	1	4
1977	Alfred Teves GmbH 3170 Gifhorn	Engineering	Servicing of production	Machine parts in containers	0,5 t	1 EGZ-I 20 Electric tractor	200	2	4
1977	Pfaff Industreimasch. GmbH 6750 Kaiserslautern	Engineering	Servicing of production	Machine parts on pallets	0,2 t	1 EGZ-I-20 Electric tractor	90	1	8
1978	Grundig AG, Werk 21 85 Nürnberg	Engineering	Servicing of production	Electrical small parts in cases	0,3 t	5 EGZ-I 20 Electric tractors	1700	27	28

Installations of Jungheinrich, West Germany (Domestic)

YEAR OF SALE	USER/SITE	BRANCH OF INDUSTRY	APPLICATION	TRANSPORT GOODS LOADING EQUIPMENT	Transport weight in tonnes	TRUCK TYPE/ TRUCK MAKE	NETWORK Length in m	NETWORK Number of switching stops	NETWORK Number of stopping points
1978	Ymos 6580 Idar-Oberstein	Engineering	Servicing of production	Machine parts on pallets roller containers	1,5 t	3 EGU-I 20 Electric pole pallet trucks	400	7	38
1978	Schott 8372 Zwiesel	Glass industry	Servicing of production	Shaped glass in skeleton containers on roller containers	0,4 t	2 EGZ-I 20 Electric tractors	400		
1979	Schott 8372 Zwiesel	Glass industry	Servicing of production	Shaped glass on pallets	0,6 t	3 EGZ-I 20 Electric tractors	900		
1979	Gesamthochschule Duisburg	Research	Supply of machine tools		1 t	3 EGP-I Special trucks	90	3	5
1979	Busch-Jäger 5880 Lüdenscheid	Electrical industry	Servicing of production	Electrical parts on pallets	0,8 t	4 EGZ-I 20 Electric tractors	880	12	35

Installations of Jungheinrich, West Germany (Domestic)

APPENDIX

YEAR OF SALE	USER/SITE	BRANCH OF INDUSTRY	APPLICATION	TRANSPORT GOODS LOADING EQUIPMENT	Transport weight in tonnes	TRUCK TYPE/ TRUCK MAKE	Length in m	NETWORK Number of switching stops	Number of stopping points
1979	Geo Gleistein 2820 Bremen	Spinning mill and cordage	Servicing of production	Cordage parts on pallets	1 t	1 EGZ-I 20 Electric tractor	350		
1979	Weco Polstermöbel 5488 Leimbach	Furniture industry	Distribution and transport in the storage area	furniture on pallets and stands	0,6 t	3 EGU-I 15 Electric pole pallet trucks	1100		
1980	Siemens AG 1000 Berlin 13	Electrical industry	Servicing of production	Electrical parts on pallets	0,8 t	2 EGZ Electric tractors	400		
1980	Emsa-Werke 4407 Emsdetten	Plastics industry	Servicing of production	Plastic parts on pallets	0,4 t	2 EGU-IR 16 Electric pole pallet trucks	650		
1980	Volkswagenwerk 3321 Salzgitter-Beddingen	Car industry	Assembly work	Engines on special pallets	0,25t	50 EGP-I Special assembly trucks	600		

Installations of Jungheinrich, West Germany (Domestic)

YEAR OF SALE	USER/SITE	BRANCH OF INDUSTRY	APPLICATION	TRANSPORT GOODS LOADING EQUIPMENT	Transport weight in tonnes	TRUCK TYPE/ TRUCK MAKE	NETWORK Length in m	NETWORK Number of switching stops	NETWORK Number of stopping points
1980	Philip Morris 8000 München	Cigarette industry	Distribution and transport in the storage area	Cigarettes on pallets	0,6 t	1 EGU-I-R Electric pole pallet truck	200		
1980	I.-A. Schnell 2354 Hohenwestedt	Food industry	Distribution and transport in the storage area	Food on roller containers	0,8 t	2 EGZ-I 50 Electric tractors	470		
1980	Purchase office of German ironmongers Wuppertal	Trade	Distribution and transport in the storage area	Small parts on pallets	0,8 t	2 EGZ Electric tractors	350		
1980	FAG Schweinfurt	Engineering	Servicing of production	Machine parts on pallets	1 t	7 special trucks with chain conveyor	250		
1981	COOP 7128 Trossingen	Food industry	Distribution and transport in storage areas	Food on rolling containers	0,4 0,8 t	8 EGA-I 7 EGU-I	1350		

Installations of Jungheinrich, West Germany (Domestic)

APPENDIX

YEAR OF SALE	USER/SITE	BRANCH OF INDUSTRY	APPLICATION	TRANSPORT GOODS LOADING EQUIPMENT	Transport weight in tonnes	TRUCK TYPE/ TRUCK MAKE	NETWORK Length in m	NETWORK Number of switching stops	NETWORK Number of stopping points
1982	COOP 3203 Sarstedt	Food industry	Distribution and storage in the storage areas	Food on rolling containers	0,4 0,8	10 EGA-I 5 EGU-I	900		
1982	Efka 7218 Trossingen	Cigarette industry	Distribution and transport in the storage areas	Cigarette papers on pallets	1,5	2 EFZ-I 20 tractor	550		
1982	Jungheinrich 2000 Hamburg	Engineering	Servicing of production	Metal parts on pallets	1	1 EGU-IR 16	300		
1982	Keiper 6760 Rockenhausen	Automobile industry	Distribution and transport in storage areas	Automobile parts on pallets	1	4 EFZ-I 50	1050		
1982	Kupplungstechnik 4440 Rheine	Engineering	Servicing of production	Metal parts in boxes and pallets	1	2 EGU-I 16	160		

Installations of Jungheinrich, West Germany (Domestic)

YEAR OF SALE	USER/SITE	BRANCH OF INDUSTRY	APPLICATION	TRANSPORT GOODS LOADING EQUIPMENT	Transport weight in tonnes	TRUCK TYPE/ TRUCK MAKE	NETWORK Length in m	Number of switching stops	Number of stopping points
1982	4 P Nikolaus 8960 Kempten		Distribution and transport in storage areas	Paper on pallets	1,6	3 EGU-IR 16	360		
1982	Siemens AG 7500 Karlsruhe	Electrical industry	Servicing of production	Electrical parts in bins	0,1	2 EGZ-IR 01	450		
1983	Apura 6502 Mainz-Kostheim	Cosmetics industry	Distribution and transport in the storage areas	Tissue paper on pallets	0,8	7 EGP-IR 10	350		
1983	Frauenhofer Inst. für Prod.-Technik und Autom. 7000 Stuttgart	Institute		Machined parts on special pallets	1	1 EGP-IR 10	50		
1983	Siemens AG 7500 Karlsruhe	Electrical industry	Servicing of production	Sorted magazines	0,1	2 EGP-I 01	340		

Installations of Jungheinrich, West Germany (Domestic)

APPENDIX

YEAR OF SALE	USER/SITE	BRANCH OF INDUSTRY	APPLICATION	TRANSPORT GOODS LOADING EQUIPMENT	Transport weight in tonnes	TRUCK TYPE/ TRUCK MAKE	NETWORK Length in m	NETWORK Number of switching stops	NETWORK Number of stopping points
1965	Swiss Air Kloten/Switzerland	Aviation	Distribution and transport in the storage area			1 EJE-HF 20 Electric pallet truck	200	–	15
1967 1975	Spinnerei an der Lorze Barr/Switzerland					1 EDZ-HF 40 Electric pole tractor	600	3	15
1968	AB Casco Kristinehamn/Sweden	Chemical industry				1 EDZ-HF 50 Electric pole tractor	530	–	9
1968	Papeteries de la Seine Nanterre/France	Paper industry				1 EDZ-HF 40 Electric pole tractor	160	–	4
1969	Pirelle Messina/Sicily/Italy	Chemical industry				1 EDZ-HF 80 Electric pole tractor	730	2	4

Installations of Jungheinrich, West Germany (Elsewhere)

YEAR OF SALE	USER/SITE	BRANCH OF INDUSTRY	APPLICATION	TRANSPORT GOODS LOADING EQUIPMENT	Transport weight in tonnes	TRUCK TYPE/ TRUCK MAKE	NETWORK Length in m	NETWORK Number of switching stops	NETWORK Number of stopping points
1969 1975	Philips Eindhoven/Holland	Electrical industry	Distribution and transport in the storage area	Pallets		6 EDZ-HF 40 Electric pole tractors	700	3	4
1970	Siemens AG Oostkamp/Belgium	Electrical industry				1 EDZ-HF 40 Electric pole tractor	300	2	5
1970 1975	Bekaert Zwevegen/Belgium	Metal industry				1 EDZ-HF 40 Electric pole tractor	350	–	10
1970	Sidac, Gent/Belgium	Chemical industry				1 EDZ-HF 20 Electric pole tractor	220	–	11
1970	Lucas, Haddenham/Great Britain	Electrical industry	Servicing of production	Electrical parts		1 EDZ-HF 100 Electric pole tractor	350	–	4

Installations of Jungheinrich, West Germany (Elsewhere)

APPENDIX 219

YEAR OF SALE	USER/SITE	BRANCH OF INDUSTRY	APPLICATION	TRANSPORT GOODS LOADING EQUIPMENT	Transport weight in tonnes	TRUCK TYPE/ TRUCK MAKE	NETWORK			
							Length in m	Number of switching stops	Number of stopping points	
1970	Spinnerei Kunz, Windisch/Switzerland	Spinning mill				2 EDZ-HF 40 Electric pole tractors	1320	5	55
1971 1974	Philips, Vienna/Laxenberg/ Austria	Electrical industry				8 EDZ-HF 40 Electric pole tractors	3000	23	51
1972	Wehkamp, Zwolle/Holland	Dispatch	Distribution and transport in the storage area	Department store articles on pallets		2 EDZ-HF 20 Electric pole tractors	600	–	10
1973	Sunlight, Olten/Switzerland	Chemical industry	Distribution and transport in the storage area			4 EDZ-HF 40 Electric pole tractors	700	15	32
1973	Weber, Aarburg/Switzerland	Spinning mill		Pallets		1 EJE-HF 20	850	4	7

Installations of Jungheinrich, West Germany (Elsewhere)

YEAR OF SALE	USER/SITE	BRANCH OF INDUSTRY	APPLICATION	TRANSPORT GOODS LOADING EQUIPMENT	Transport weight in tonnes	TRUCK TYPE/ TRUCK MAKE	NETWORK Length in m	Number of switching stops	Number of switching stopping points
1973	Möbel-Pfister, Suhr/Switzerland	Furniture industry	Distribution and transport in the storage area	Furniture on special trailers		20 EFZ-HF 80 Electric driver seated tractors	8000	44	126
1973	Sulzer, Zuchwil/Switzerland	Engineering	Servicing of production	Machine parts in special holders		3 Mini EJZ-HF Electric tractors	600	4	40
1973 1975	Sulzer, Zuchwil/Switzerland	Engineering	Servicing of production	Machine parts on special holders		5 EDZ-HF 40 Electric pole tractors	3000	36	62
1974	Schöller Albers, Schaffhausen/ Switzerland	Textile industry				1 EDZ-HF 40 Electric driver seated tractor	400	–	3
1974	AB Hernia, Norrköping/Sweden	Chemical industry	Servicing of production	Adhesives carried on trailers		1 EFZ-HF 80 Electric driver seated tractor	640	3	9

Installations of Jungheinrich, West Germany (Elsewhere)

APPENDIX

YEAR OF SALE	USER/SITE	BRANCH OF INDUSTRY	APPLICATION	TRANSPORT GOODS LOADING EQUIPMENT	Transport weight in tonnes	TRUCK TYPE/ TRUCK MAKE	NETWORK		
							Length in m	Number of switching stops	Number of stopping points
1974	Borckenstein, Neudau/Austria	Spinning mill	Servicing of production			1 EDZ-HF 50 Electric pole tractor	580	18	manual
1974	Cartonnerie de la Rochette, de l'Hermitage, Rochette/France	Paper industry	Servicing of production	Paper on trailers		1 EDZ-HF 120 Electric pole tractor	150	2	10
1974	Isoverbel, Etten Leux/Holland	Chemical industry	Servicing of production	Insulation material on trailers		5 EDZ-HF 20 Electric pole tractors	700	5	6
1974	Papeteries, Haseldonck/Belgium	Paper industry	Servicing of production			1 EDZ-HF 50 Electric pole tractor	300	–	2
1975	ITA, Serval le Chateau/France	Furniture industry		Furniture on special trailers		1 EDZ-HF 20 Electric pole tractor	200	3	10

Installations of Jungheinrich, West Germany (Elsewhere)

YEAR OF SALE	USER/SITE	BRANCH OF INDUSTRY	APPLICATION	TRANSPORT GOODS LOADING EQUIPMENT	Transport weight in tonnes	TRUCK TYPE/ TRUCK MAKE	NETWORK		
							Length in m	Number of switching stops	Number of stopping points
1975	Anic PT 1, Ottana Nouro/Sardinia/ Italy	Textile industry	Servicing of production			5 EDZ-HF 20 Electric pole tractors	800	5	8
1975	Anic PT 2, Ottana/Nouro/Sardinia/ Italy	Textile industry	Servicing of production			4 EDZ-HF 80-Sonder Electric pole tractors	2000	31	43
1975	Contonificio di Conegliano/Conegliano Pordenone/Italy	Textile industry				2 EDZ-HF 20 Electric pole tractors	100	–	3
1975	3 M Italiana, Savona/Italy	Chemical industry	Servicing of production	Photographs		2 EDZ-HF 20 Electric pole tractors	120	2	14
1976	Zignago S. Margherita, Portogruaro/Venice/ Italy	Glass industry	Servicing of production	Glass on pallets		2 EGZ-I 50 Electric tractors	300	1	12

Installations of Jungheinrich, West Germany (Elsewhere)

APPENDIX

YEAR OF SALE	USER/SITE	BRANCH OF INDUSTRY	APPLICATION	TRANSPORT GOODS LOADING EQUIPMENT	Transport weight in tonnes	TRUCK TYPE/ TRUCK MAKE	NETWORK Length in m	NETWORK Number of switching stops	NETWORK Number of stopping points
1976	SCM Società Costruzione Macchine, Rimini/Italy	Iron and metal industry	Servicing of production	Machine parts on trailers		1 EFZ-I 150 Electric driver seated tractor	370	–	4
1976	Ciba Geigy, Basel/Stein/ Switzerland	Chemical industry	Servicing of production	Pharmaceutical products		3 EFZ-I 100 Electric driver seated tractors	530	–	2
1976	Sulzer, Winterthur/Switzerland	Engineering	Servicing of production	Machine parts		1 EDZ-HF 20 Electric pole tractor	120	–	5
1977	Montefibre, Gruppo Montedison, Acerra (Naples) Italy	Textile industry	Servicing of production	Spinning spools on special trailers		2 EGZ-IR 50-Special electric tractors	500		
1977	Mifattura, Contoniera Merideonale, Fratte (Salerno) Italy	Textile industry	Servicing of production	Spinning spools on special trailers		4 EGZ-I 20 Electric tractors	840	6	32

Installations of Jungheinrich, West Germany (Elsewhere)

YEAR OF SALE	USER/SITE	BRANCH OF INDUSTRY	APPLICATION	TRANSPORT GOODS LOADING EQUIPMENT	Transport weight in tonnes	TRUCK TYPE/ TRUCK MAKE	NETWORK		
							Length in m	Number of switching stops	Number of stopping points
1977	Illochroma, Brussels/Belgium	Paper industry	Servicing of production	Paper on pallets		1 EGU-IR Electric pole pallet truck	440	30	
1977	Elida Gibbs Leeds/Great Britain	Chemical industry	Servicing of production	Cosmetic articles on pallets		3 EGZ-I 100 Electric tractors	350	—	14
1977	Papierfabrik Laakirchen, Laakirchen/Austria	Paper industry	Servicing of production	Paper on pallets		5 EGS-I	300	1	52
1977	John Waddington Leeds/Great Britain	Plastics industry	Production transport	Pallets		2 EFZ/ 50	175	2	4
1977	Volvo Skövde/Sweden	Automobile industry		Pallets		1 EFZ I 150	360		

Installations of Jungheinrich, West Germany (Elsewhere)

APPENDIX

YEAR OF SALE	USER/SITE	BRANCH of INDUSTRY	APPLICATION	TRANSPORT GOODS LOADING EQUIPMENT	Transport weight in tonnes	TRUCK TYPE/ TRUCK MAKE	NETWORK Length in m	NETWORK Number of switching stops	NETWORK Number of stopping points
1977	Elida Gibbs Leeds/England	Chemical industry	Warehousing	4 pallets and roller hangers	1 t	3 EGZ I 100	440		14
1975 1980	Beghin Hondouville/France	Paper industry	Servicing of production	Pallets	6 t	2 EDZ HF 50 1 EGZ I 50	100	1	15
1978	Reeds Corrugated Nottingham/England	Paper industry	Transport to production station	4m long cartons with roller hangers		2 EFZ I 100	750		6
1978	Birds Eye Food Liverpool/England	Food industry	Servicing of production	Pallets		2 EGS 10	180	1	12
1978	Cloetta Ljungsbro/Sweden	Food industry	Servicing of production	Special pallets		2 EGU IR	450	36	

Installations of Jungheinrich, West Germany (Elsewhere)

| YEAR OF SALE | USER/SITE | BRANCH OF INDUSTRY | APPLICATION | TRANSPORT GOODS LOADING EQUIPMENT | Transport weight in tonnes | TRUCK TYPE/ TRUCK MAKE | NETWORK |||
							Length in m	Number of switching stops	Number of stopping points
1978 1981	Fraymon S.A. Madrid/Spain	Automobile industry	Servicing of production	Pallets	3 t	3 EFZ I 50	633	4	7
1979	Triumph International Swindon/England	Textile industry	Transport to production station	3 pallets with special hangers		1 EGZ 50	200	17	14
1979	Santagata Rocchetta e Croce/Italy	Food industry	Servicing of production	Pallets with mineral water		3 EFZ I 50	360	5	8
1979	Geberit AG Lona/Switzerland	Sanitary ware	Servicing of production		2 t	9 EGW I	1600		
1979	Swiss Railway and air terminal Kloten/Switzerland	Station	Transport			5 EDZ HF	760		

Installations of Jungheinrich, West Germany (Elsewhere)

YEAR OF SALE	USER/SITE	BRANCH OF INDUSTRY	APPLICATION	TRANSPORT GOODS LOADING EQUIPMENT	Transport weight in tonnes	TRUCK TYPE/ TRUCK MAKE	NETWORK			
							Length in m	Number of switching stops	Number of stopping points
1979	Matcene Belgium	Food industry	Servicing of production	Pallets		2 EGU I 50	300		4
1979	Agfa-Gevaert Belgium	Photographic industry	Servicing of production		5 t	2 EGU I 50	300		4
1980	Philips Vendre Belgium		Servicing of production	Pallets	5 t	3 EFZ I 50	730		7
1980	Rhone Poulenc Fibres Arras/France	Spinning industry	Servicing of production	Spinning reels on special hangers	0,6 t 2 t	6 EGU IR 06 3 EFZ I 20	1000		
1980	Carticre del Garda Riva del Garda/Italy	Paper industry	Servicing of production			5 EGU IR	300	20	16

Installations of Jungheinrich, West Germany (Elsewhere)

YEAR OF SALE	USER/SITE	BRANCH OF INDUSTRY	APPLICATION	TRANSPORT GOODS LOADING EQUIPMENT	Transport weight in tonnes	TRUCK TYPE/ TRUCK MAKE	NETWORK		
							Length in m	Number of switching stops	Number of stopping points
1980	Artsana Fino Mornasco/Italy	Baby and children's articles	Servicing of production			3 EGZ I 20			
1980	Scania Zwolle/Holland	Automobile industry	Transport of motors	Special carriers	1,2 t	1 EGU I	175	1	4
1980	Daells Varehus Kopenhagen/Dänemark		Servicing of production			2 EGZ I 50	440		5
1980	Gebr. Sulzer AG Zuchwil/Switzerland	Engineering	Assembly			8 EMS I	140		
1981	Eidg. Munitionsfabrik Altdorf/Switzerland	Army	Transport in the laboratory	Pallets and hand pallets	5 t	2 EDZ HF	320		

Installations of Jungheinrich, West Germany (Elsewhere)

APPENDIX

YEAR OF SALE	USER/SITE	BRANCH OF INDUSTRY	APPLICATION	TRANSPORT GOODS LOADING EQUIPMENT	Transport weight in tonnes	TRUCK TYPE/ TRUCK MAKE	NETWORK		
							Length in m	Number of switching stops	Number of stopping points
1981	Leglertex, Ponte s. Pietro Wahlendorf/ Switzerland	Textile industry	Supply and retrieval for production	Textile rolls		4 EGW I			
1982	Bristol Allard Lognes/France	Pharmaceutical industry	Supply and retrieval for production	Pallets	2 t	1 EFZ I 20	300	2	7
1982	SNIAS (Airbus) Nantes/France	Aerospace industry	Supply and retrieval for production	Pallets	2 t	2 EGZ I 20	770		15
1982	GEC. Traction Manchester/England	Engineering	Warehouse transport	Pallets		1 EGP	160	6	6
1982	Leglertex P.te S. Pietro/Italy	Textile industry	Supply and retrieval for production	Cloth rolls		5 EGW I 20 with special load pick-up device	550	10	35

Installations of Jungheinrich, West Germany (Elsewhere)

YEAR OF SALE	USER/SITE	BRANCH OF INDUSTRY	APPLICATION	TRANSPORT GOODS LOADING EQUIPMENT	Transport weight in tonnes	TRUCK TYPE/ TRUCK MAKE	NETWORK Length in m	Number of switching stops	Number of stopping points
1982	VM Cento (FE)/Italy	Engineering	Supply and retrieval for production	Motors on special pallets		2 EGZ I 20 RB	300	3	56
1982	Weisbrod-Zürrer AG Mettmenstetten/ Switzerland	Paper industry	Supply and retrieval for production	Pallets		3 EGU IR	500		
1982	Suchard AG Neuchatel/Switzerland	Food industry	Order picking	Chocolate on pallets		2 EGU I KmS			
1982	Bekeart Waregem Belgium	Textile industry	Supply and retrieval for production	Spinning reels		1 ESZ I 10	500		
1983	Renault Swindon/England	Automobile industry	Distribution	Three pallets on hand trailer		6 EGZ 50	600	9	8

Installations of Jungheinrich, West Germany (Elsewhere)

APPENDIX

YEAR OF SALE	USER/SITE	BRANCH OF INDUSTRY	APPLICATION	TRANSPORT GOODS LOADING EQUIPMENT	Transport weight in tonnes	TRUCK TYPE/ TRUCK MAKE	NETWORK		
							Length in m	Number of switching stops	Number of stopping points
1983	Ford Motor Dagenham/England	Automobile industry	Transport in production	Two special pallets on elevating table		14 EGP	3200	12	8
1983	Renault Italia S. Cdombano Lambro/ Italy	Automobile industry	Supply and retrieval for production	Spare parts on pallets and boxes		2 EGZ I 50	600	6	6
1983	Vrumona Bunnik/Holland	Food industry		Transport of drinks	2,2 t	2 EGZ I 15	510	2	3
1983	Escher SA Genf/Switzerland	Food industry		Wine on pallets		2 EGW I			

Installations of Jungheinrich, West Germany (Elsewhere)

YEAR OF SALE	USER/SITE	BRANCH OF INDUSTRY	APPLICATION	TRANSPORT GOODS LOADING EQUIPMENT	Transport weight in tonnes	TRUCK TYPE/ TRUCK MAKE	NETWORK Length in m	NETWORK Number of switching stops	NETWORK Number of switching stopping points
1976	University of Dortmund	Research	Servicing of a high-bay warehouse	Pallets	0,5 t	1 EFZ HP 01 Electric tractor	80		
1980	Robert Bosch GmbH 7000 Stuttgart	Electrical industry	Servicing of production	Electrical parts on pallets	0,5 t	1 EFZ-SPT 02 Electric tractor	100		
1980	Robert Bosch GmbH 7000 Stuttgart-Feuerbach	Automobile industry		Pallets	0,4 0,8	2 EFZ-SPT 01	300	10	
1981	Robert Bosch GmbH 7000 Stuttgart-Feuerbach	Automobile industry		Reflectors on pallets in boxes	0,5	1 TSPT	120		
1982	Robert Bosch GmbH 7000 Stuttgart-Feuerbach	Automobile industry	Distribution	Reflectors on pallets in boxes	0,5	1 EGN 01	180		

Installations of PHB-Babcock

APPENDIX

YEAR OF SALE	USER/SITE	BRANCH OF INDUSTRY	APPLICATION	TRANSPORT GOODS LOADING EQUIPMENT	Transport weight in tonnes	TRUCK TYPE/ TRUCK MAKE	NETWORK		
							Length in m	Number of switching stops	Number of stopping points
1982	VFM/MBB 2800 Bremen	Aerospace industry		Sheet metal parts	0,5 0,25	4 EGH 01	1000		
1983	Zellweger Uster AG Uster		Servicing of production	Special pallets	0,4 t 0,8 t	2 EFZ-SPT 01 Electric trucks	300	10	

Installations of PHB-Babcock

YEAR OF SALE	USER/SITE	BRANCH OF INDUSTRY	APPLICATION	TRANSPORT GOODS LOADING EQUIPMENT	Transport weight in tonnes	TRUCK TYPE/ TRUCK MAKE	NETWORK Length in m	NETWORK Number of switching stops	NETWORK Number of stopping points
1978	Becker Autoradiowerk Karlsbad	Electrical industry	Servicing of production	Car radios, accessories on pallets	0,5 t	1 Electric truck	210		5
1979	Bosch-Lichtwerke Stuttgart-Feuerbach	Electrical industry	Servicing of production	Electrical small parts in containers & skeleton containers	0,5 t	1 Electric truck	160		
1981	Bosch GmbH 7000 Stuttgart-Feuerbach West Germany	Automobile industry		Small parts for reflectors	0,25	1 Telelift-Transcar	140		
1981	Berg. Kaserne 4000 Düsseldorf-Hubbelrath West Germany		Special transport	Prepared food in containers	0,3	3 Telelift/Transcar	170		
1981	J2T Video GmbH 1000 Berlin West Germany	Electrical industry	Supply and retrieval for production	Assembly material in special containers	0,15	10 Telelift-Transcar	520		

Installations of Telelift, West Germany

APPENDIX

YEAR OF SALE	USER/SITE	BRANCH OF INDUSTRY	APPLICATION	TRANSPORT GOODS LOADING EQUIPMENT	Transport weight in tonnes	TRUCK TYPE/ TRUCK MAKE	NETWORK Length in m	Number of switching stops	Number of stopping points
1982	Bosch-Lichtwerke 7000 Stuttgart-Feuerbach West Germany	Electrical industry	Supply and retrieval for production	Window wipers	0,5	2 Telelift-Transcar			

Installations of Telelift, West Germany

YEAR OF SALE	USER/SITE	BRANCH OF INDUSTRY	APPLICATION	TRANSPORT GOODS LOADING EQUIPMENT	Transport weight in tonnes	TRUCK TYPE/ TRUCK MAKE	NETWORK Length in m	Number of switching stops	Number of stopping points
1973	Volvo-Kalmarverken, Sweden	Car industry	Assembly work	Bodywork		1 Low carrier special type			
1973	Volvo-Kalmarverken Sweden	Car industry	Assembly work	Bodywork		1 High carrier special type			
1975	Volvo-Europa NV. Gent, Belgium	Car industry	Assembly work	Bodywork		50 Low carrier special type			
1975	Volvo Europa NV. Gent, Belgium	Car industry	Assembly work	Bodywork		4 High carrier special type			
1975	Volvo-Torslandaverken Sweden	Car industry	Servicing of production	Cars		3 Telecarrier side loaders			

Installations of Tellus, Sweden

APPENDIX

| YEAR OF SALE | USER/SITE | BRANCH OF INDUSTRY | APPLICATION | TRANSPORT GOODS LOADING EQUIPMENT | Transport weight in tonnes | TRUCK TYPE/ TRUCK MAKE | NETWORK |||
							Length in m	Number of switching stops	Number of switching stopping points
1976	Tetra Pak Lund, Sweden	Paper industry	Servicing of production	Paper rolls		20 Telecarrier electric pallet trucks			
1976	Volvo-Torslandaverken Sweden	Car industry	Assembly work	Bodyworks		2 Telecarrier special type			
1976	Grycksbo Pappersbruck, Sweden	Paper industry	Servicing of production	Paper rolls		2 Telecarrier electric pallet trucks			
1976	Volvo-Tuve Sweden	Car industry	Servicing of production	Car parts on pallets		15 Telecarrier special type			
1976	Tetra Pak Lund/Sweden	Paper industry	Servicing of production	Boards		1 Telecarrier Electric pallet truck			

Installations of Tellus, Sweden

YEAR OF SALE	USER/SITE	BRANCH OF INDUSTRY	APPLICATION	TRANSPORT GOODS LOADING EQUIPMENT	Transport weight in tonnes	TRUCK TYPE/ TRUCK MAKE	NETWORK Length in m	Number of switching stops	Number of stopping points
1977	Valco Oy, Imatra Finland	Electrical industry	Assembly work	Television tubes		30 Telecarrier special type			
1977	Tetra Pak Lund/Sweden	Paper industry	Servicing of production	Paper rolls		4 Telecarrier Electric pallet trucks			
1978	De Lorean Motor Company Dunmurry Northern Ireland	Engineering	Assembly work	Engines		1 Low carrier special type			
1978	Langereds Färg, Karlstad/Sweden			Pallets		1 Telecarrier electric tractor			
1979	Svenska Gods-Centraler ASG, Stockholm/Sweden	Food industry	Distribution and transport in the storage area	Foodstuffs and luxury goods on pallets		2 Telecarrier electric tractors			

Installations of Tellus, Sweden

APPENDIX

YEAR OF SALE	USER/SITE	BRANCH OF INDUSTRY	APPLICATION	TRANSPORT GOODS LOADING EQUIPMENT	Transport weight in tonnes	TRUCK TYPE/ TRUCK MAKE	NETWORK			
							Length in m	Number of switching stops	Number of switching stopping points
1979	N.V. Lithorex S.A. Erembodegem/Belgium	Paper industry	Servicing of production	Paper rolls		1 Telecarrier electric pallet truck			
1979	Vin- & Spritcentralen Stockholm/Sweden	Drinks trade		Drinks on pallets		7 Telecarrier Electric tractors			
1979	De Lorean Motor Company Dunmurry Northern Ireland	Engineering	Assembly work	Engines		1 Lowcarrier special type			
1980	De Lorean Motor Company Dunmurry Northern Ireland	Engineering	Assembly work	Engines		35 Lowcarrier special type			
1980	Saab-Scania Oakarshamn/Sweden	Truck construction	Servicing of production	Lorry cab bodyworks		9 Telecarrier Special type			

Installations of Tellus, Sweden

YEAR OF SALE	USER/SITE	BRANCH OF INDUSTRY	APPLICATION	TRANSPORT GOODS LOADING EQUIPMENT	Transport weight in tonnes	TRUCK TYPE/ TRUCK MAKE	NETWORK		
							Length in m	Number of switching stops	Number of stopping points
1980	Volvo-Olofströmaverken Sweden	Car industry	Servicing of production	Metal pallets		7 Telecarrier Special type			
1980	Tetra Pak International Sweden	Paper industry	Servicing of production	Paper rolls		80 Telecarrier Electric pallet trucks			
1980	Iggesunds Bruk Sweden	Paper industry	Servicing of production	Paper rolls		2 Telecarrier Electric pallet trucks			
1981	Saab-Scania Oskarshamn/Sweden	Automobile industry		Truck cabs		1			
1981	Tetra Pak Dijon/France	Paper industry		Paper rolls		10			

Installations of Tellus, Sweden

APPENDIX 241

YEAR OF SALE	USER/SITE	BRANCH OF INDUSTRY	APPLICATION	TRANSPORT GOODS LOADING EQUIPMENT	Transport weight in tonnes	TRUCK TYPE/ TRUCK MAKE	NETWORK Length in m	Number of switching stops	Number of stopping points
1981	Tetra Pak Moerdijk/Holland	Paper industry		Paper rolls	10				
1982	Tetra Pak Arganda del Ray/Spain	Paper industry		Paper rolls	10				
1982	Tetra Pak Rubiera/Italy	Paper industry		Paper rolls	10				
1982	Tetra Pak Rubiera	Paper industry		Pallets	3				
1983	Tetra Pak Arganda de Ray/Spain	Paper industry		Pallets	2				

Installations of Tellus, Sweden

YEAR OF SALE	USER/SITE	BRANCH OF INDUSTRY	APPLICATION	TRANSPORT GOODS LOADING EQUIPMENT	Transport weight in tonnes	TRUCK TYPE/ TRUCK MAKE	NETWORK Length in m	Number of switching stops	Number of stopping points
1983	Tetra Pak Romont/Swiss	Paper industry		Paper rolls	8				
1983	Tetra Pak Romont/Swiss	Paper industry		Pallets	1				
1983	Posten Tomteboda Stockholm/Sweden	Post Office		Pallets	0,3	50 Forktrucks			
1983	Snia Viscosa Italy (Anlage in der UdSSR)	Chemical industry		Glass fibre in rolls	1,0	131 Trucks with rollerbed			
1983	Leykam Muerztaler Gratkorn/Austria	Paper industry		Paper rolls	3,0	9 Special forktrucks			

Installations of Tellus, Sweden

APPENDIX
243

YEAR OF SALE	USER/SITE	BRANCH OF INDUSTRY	APPLICATION	TRANSPORT GOODS LOADING EQUIPMENT	Transport weight in tonnes	TRUCK TYPE/ TRUCK MAKE	Length in m	NETWORK Number of switching stops	Number of stopping points
1983	Giddings & Lewis Corp. Anderson & Strathclyde Motherwell Scotland in Cooperation with Conco-Tellus					1 Fork lift truck			
1983	SKF Gothenburg/Sweden	Engineering		Special pallets	1,0	1 Fork lift truck			
1983	Tetra Pak Lund/Sweden	Paper industry		Paper rolls		15 Trucks			

Installations of Tellus, Sweden

YEAR OF SALE	USER/SITE	BRANCH OF INDUSTRY	APPLICATION	TRANSPORT GOODS LOADING EQUIPMENT	Transport weight in tonnes	TRUCK TYPE/ TRUCK MAKE	NETWORK Length in m	Number of switching stops	Number of stopping points
1974	Volvo/Skörde Sweden	Car industry	Assembly work	Engines		360 T-carriers Assembly trucks			
1976	Volvo/Venlo Sweden	Car industry	Assembly work	Cars		54 low-carriers Skid vehicle			
1977	Volvo/Umea Sweden	Car industry	Assembly work	Bodywork		46 low-carriers Skid vehicle			
1977	Volvo/Torslanda Sweden	Car industry	Servicing of production	Cars		1-side-loader			
1978	Volvo/Tuve Sweden	Car industry	Servicing of production	Car parts		16-carriers			

Installations of ACS, Sweden

YEAR OF SALE	USER/SITE	BRANCH OF INDUSTRY	APPLICATION	TRANSPORT GOODS LOADING EQUIPMENT	Transport weight in tonnes	TRUCK TYPE/ TRUCK MAKE	NETWORK Length in m	NETWORK Number of switching stops	NETWORK Number of stopping points
1978	Atla/Linköping Sweden	Food industry	Distribution and transport in the storage area	Milk products		8-3-fork-carriers Electric pallet truck			
1979	VVA Gütersloh	Book trade	Commissioning	Books on pallets		22 carriers	1500	40	80
1980	Kutel, Essen	Food industry	Servicing of production	Milk products		7 carriers			
1980	Volvo Torslanda Sweden	Car industry	Assembly work	Bodyworks		22 carriers			
1981	Mölnlycke Gothenburg/Sweden		Distribution	Pallets		3 Carriers with rollertable			

Installations of ACS, Sweden

YEAR OF SALE	USER/SITE	BRANCH OF INDUSTRY	APPLICATION	TRANSPORT GOODS LOADING EQUIPMENT	Transport weight in tonnes	TRUCK TYPE/ TRUCK MAKE	NETWORK Length in m	Number of stops	Number of switching stops	Number of stopping points
1981	Scania Zwolle/Holland	Automobile industry	Distribution	Motors and trucks		1 Carrier				
1981	Volvo Comp. Skövde/Sweden	Automobile industry	Servicing NC machines	Gearbox housings		2 TP201 (carrier)				
1981	Volvo Vomp. Sköde/Sweden	Automobile industry	Supply and retrieval in production	Crankshafts		3 TP201 (carrier)				
1982	Volvo Truck Ostakker/Belgium	Automobile industry	Assembly			6 Carriers with trolleys				
1982	Arla Kallhäll/Sweden	Food industry	Distribution	Baskets		4 Carrier with double roller table				

Installations of ACS, Sweden

APPENDIX

YEAR OF SALE	USER/SITE	BRANCH OF INDUSTRY	APPLICATION	TRANSPORT GOODS LOADING EQUIPMENT	Transport weight in tonnes	TRUCK TYPE/ TRUCK MAKE	Length in m	NETWORK Number of switching stops	Number of stopping points
1982	Arla Örebro/Sweden	Food industry	Distribution	Baskets		1 Carrier with double roller table			
1982	Saab Trollhätten/Sweden	Automobile industry	Assembly	Motors and gearboxes with exhaust		35 Carriers			
1982	Saab Trollhätten/Sweden	Automobile industry	Transport	Car bodies		4 TP207 (carriers)			
1982	Volvo Comp. Skövde/Sweden	Automobile industry	Assembly	Motors		40 Carriers			
1982	Volvo Comp. Skövde/Sweden	Automobile industry	Distribution	Pallets		16 Carriers			

Installations of ACS, Sweden

YEAR OF SALE	USER/SITE	BRANCH OF INDUSTRY	APPLICATION	TRANSPORT GOODS LOADING EQUIPMENT	Transport weight in tonnes	TRUCK TYPE/ TRUCK MAKE	NETWORK Length in m	Number of switching stops	Number of stopping points
1982	W. Karmann GmbH Osnabrück/West-Germany	Automobile industry	Supply and retrieval in assembly	Bodies on pallets	0,5	7 TPO 207 (assembly carriers)	150		

Installations of ACS, Sweden

| YEAR OF SALE | USER/SITE | BRANCH OF INDUSTRY | APPLICATION | TRANSPORT GOODS LOADING EQUIPMENT | Transport weight in tonnes | TRUCK TYPE/ TRUCK MAKE | NETWORK |||
							Length in m	Number of switching stops	Number of stopping points
1978	Coop Schwaben Neuhausen	Food industry	Commissioning	Food on roller containers	0,5 t	12-BTALL-1000 Electric pallet trucks	900		
1978	OVD Trading organisations Oslo/Norway	Distribution centre	Distribution and transport in the storage area	Goods on pallets		75 BT ALL 1000 Electric pallet trucks			
1978	IKEA Almhult Sweden	Furniture industry	Distribution and transport in the storage area		0,7 t	9 BT ALC 1000 Skid units			
1979	Pharmaci Upsalla/Sweden					1 ALL 1000			
1979	Apotekarnas Fellesinköp/Norway					2 ALL 1000			

Installations of BT, Sweden

250 AUTOMATED GUIDED VEHICLES

YEAR OF SALE	USER/SITE	BRANCH OF INDUSTRY	APPLICATION	TRANSPORT GOODS LOADING EQUIPMENT	Transport weight in tonnes	TRUCK TYPE/ TRUCK MAKE	NETWORK Length in m	Number of switching stops	Number of switching stopping points
1980	Esselte Stockholm/Sweden					8 ALL 1000			
1981	Slipmaterial Naxos Västervik/Sweden					1 ALL 1000			
1981	DAGAB Stockholm/Sweden					4 APM 700			
1981	Frank Mohn Norway					2 APM 700			
1981	Volvo Färgelanda/Sweden	Automobile industry				3 APM 700			

Installations of BT, Sweden

APPENDIX   251

| YEAR OF SALE | USER/SITE | BRANCH OF INDUSTRY | APPLICATION | TRANSPORT GOODS LOADING EQUIPMENT | Transport weight in tonnes | TRUCK TYPE/ TRUCK MAKE | NETWORK |||
							Length in m	Number of switching stops	Number of stopping points
1981	KF Riskslager Bro/Sweden					7 ALL 1000			
1981	Elektolux Torsvik/Sweden	Electrical industry				4 APM 1000 3 APM 700 (1982)			
1982	Elektrolux Västervik/Sweden	Electrical industry				3 ALL 1000			
1982	Televerket Nässjo-Sweden					3 AHL			
1982	Sundsvalle Verksstäd					1 ALL			

Installations of BT, Sweden

252 AUTOMATED GUIDED VEHICLES

| YEAR OF SALE | USER/SITE | BRANCH OF INDUSTRY | APPLICATION | TRANSPORT GOODS LOADING EQUIPMENT | Transport weight in tonnes | TRUCK TYPE/ TRUCK MAKE | NETWORK |||
							Length in m	Number of switching stops	Number of stopping points
1982	IVF Stöckholm/Sweden					1 AHL			
1982	COOP Käsecentrum West-Germany	Food industry				2 ALL			
1982	ICI Great Britain	Electrical industry				3 ALL			
1982	Otto Nielsen Denmark					2 APM			
1982	BT Intertransport AS Denmark					2 APM			

Installations of BT, Sweden

APPENDIX 253

YEAR OF SALE	USER/SITE	BRANCH OF INDUSTRY	APPLICATION	TRANSPORT GOODS LOADING EQUIPMENT	Transport weight in tonnes	TRUCK TYPE/ TRUCK MAKE	NETWORK Length in m	Number of switching stops	Number of stopping points
1982	Kssko Finland					1 ALL 1000			
1982	Troodie Steffenstorp Sweden					3 APM			
1982	3M England					1 ALL			
1983	Saab Scania Sweden	Automobile industry				4 ALL 1000			
1983	Husqvarna Sweden	Engineering				3 APM 700			

Installations of BT, Sweden

YEAR OF SALE	USER/SITE	BRANCH OF INDUSTRY	APPLICATION	TRANSPORT GOODS LOADING EQUIPMENT	Transport weight in tonnes	TRUCK TYPE/ TRUCK MAKE	NETWORK Length in m	Number of stops	Number of switching stops	Number of stopping points
1983	Black & Decker Great Britain					7 ALL 1000				

Installations of BT, Sweden

APPENDIX 255

YEAR OF SALE	USER/SITE	BRANCH OF INDUSTRY	APPLICATION	TRANSPORT GOODS LOADING EQUIPMENT	Transport weight in tonnes	TRUCK TYPE/ TRUCK MAKE	NETWORK Length in m	Number of switching stops	Number of stopping points
	Citroen France	Car industry	Assembly work			Approx. 30 assembly trucks			
	Peugeot France	Car industry	Assembly work			Approx. 30 assembly trucks			

Installations of CFC, France

YEAR OF SALE	USER/SITE	BRANCH OF INDUSTRY	APPLICATION	TRANSPORT GOODS LOADING EQUIPMENT	Transport weight in tonnes	TRUCK TYPE/ TRUCK MAKE	NETWORK Length in m	NETWORK Number of switching stops	NETWORK Number of stopping points
1973	Michelin France	Chemical industry	Servicing of production	Tyres	1 t	6 TEZE Tractors	1500		
1974	Moet & Chandon Eperny/France	Drinks industry	Distribution and transport in the storage area	Drinks	2,5 t	1 TEZE Tractor			
1974	Roissy en France France	Airport	Luggage transport	Luggage	2,5 t	60 TEZE Tractors	5000		
1975	Eurodif Werk Tricastin/France	Nuclear industry	Automatic installation of compressors	Compressors	12 t	18 Special-tractors			
1976	Citroen France	Car industry	Servicing of production	Bodywork parts		1 Automatic truck			

Installations of Saxby, France

APPENDIX 257

| YEAR OF SALE | USER/SITE | BRANCH OF INDUSTRY | APPLICATION | TRANSPORT GOODS LOADING EQUIPMENT | Transport weight in tonnes | TRUCK TYPE/ TRUCK MAKE | NETWORK |||
							Length in m	Number of switching stops	Number of stopping points
1978	Peugeot Lille France	Car industry	Assembly work	Engines	0,4 t	59 Automatic trucks			
1978	Francaise de mechanique Douvrin France	Engineering	Assembly work	Engines	0,4 t	30 Automatic trucks			
1980	Peugeot Cycles France	Engineering	Assembly work	Bicycles	0,4 t	5 Automatic trucks			
1980	Soc. Mosellane De Mechanique France	Engineering	Servicing of production	Raw materials, semi-manufactured and finished goods	0,4 t	7 Automatic trucks			
1980	C.E.G.F. Loudeac/France	Nuclear industry	Transport in the storage area	Radioactive waste product	1,2 t	3 Automatic trucks			

Installations of Saxby, France

YEAR OF SALE	USER/SITE	BRANCH OF INDUSTRY	APPLICATION	TRANSPORT GOODS LOADING EQUIPMENT	Transport weight in tonnes	TRUCK TYPE/ TRUCK MAKE	NETWORK Length in m	Number of switching stops	Number of stopping points
1980	Doel nuclear power station/Belgium	Nuclear industry	Transport in the storage area	Radioactive waste products	1,2 t	3 Automatic trucks			
1981	Talbot, Poissy France	Car industry	Servicing of products	Engines	0,4 t	12 Automatic trucks			
1981	Laboratoires Delagrance, Chilly-Mazaren France	Chemical industry	Distribution and transport in the storage area	Pallets	0,4 t	3 Automatic trucks			
1981	RNUR Cléon/France	Automobile industry	Servicing of production	Single parts	0,4 t	13 Automatic trucks			
1982	RNUR Cléon/France	Automobile industry	Supply and retrieval in production	Single items		15 Automatic vehicles	5000	40	120

Installations of Saxby, France

APPENDIX

YEAR OF SALE	USER/SITE	BRANCH OF INDUSTRY	APPLICATION	TRANSPORT GOODS LOADING EQUIPMENT	Transport weight in tonnes	TRUCK TYPE/ TRUCK MAKE	NETWORK Length in m	NETWORK Number of switching stops	NETWORK Number of stopping points
1982	RNUR Cléon/France	Automobile industry	Supply and retrieval in production	Containers		40 Automatic vehicles	600	50	60
1983	Fasa Séville/Spain	Automobile industry	Supply and retrieval in production			22 Automatic vehicles	900	80	70
1983	Verreries du PUY de DOME Puy Guillaume/France	Food industry	Warehousing	Bottles in pallets		4 Automatic vehicles	400	12	11
1983	NOBEL Honfleur	Chemical industry	Transportation			3 Automatic vehicles	2000	2	14

Installations of Saxby, France

YEAR OF SALE	USER/SITE	BRANCH OF INDUSTRY	APPLICATION	TRANSPORT GOODS LOADING EQUIPMENT	Transport weight in tonnes	TRUCK TYPE/ TRUCK MAKE	NETWORK Length in m	NETWORK Number of switching stops	NETWORK Number of stopping points
1974	Volvo, Kalmar/Sweden	Car industry	Assembly work	Cars		250 Robocarriers			
1975	Fiat, Mirafiori, Turin/ Italy	Car industry	Assembly work	Chassis on assembly pallet		35 Robocarriers			
1977	Sofim, Foggia/Italy	Car industry	Servicing of production	Engines		11 Robotrailers			
1977	Citroen-Hispania, Vigo, Spain	Car industry	Assembly work	Engines		4 Robotrailers			
1977	MBB, Augsburg/ Federal Republic of Germany	Aircraft construction	Servicing of production	Precision workpieces on pallets	2,6 t	2 Robocarriers	800	60	45

Installations of Digitron, Switzerland

APPENDIX

YEAR OF SALE	USER/SITE	BRANCH OF INDUSTRY	APPLICATION	TRANSPORT GOODS LOADING EQUIPMENT	Transport weight in tonnes	TRUCK TYPE/ TRUCK MAKE	NETWORK Length in m	Number of switching stops	Number of stopping points
1977	Shell U.K. Oil, Haven, Great Britain	Chemical industry	Servicing of production	Lubricants in mix containers		4 Robotrailers			
1978	Saab-Scania, Trollhättan Sweden	Truck assembly	Servicing of production	Bodyworks		4 Robocarriers			
1978	Fiat Rivalta, Turin/Italy	Car industry	Assembly work	Bodyworks		54 Robocarriers			
1978	Fiat Cassino, Monte Cassino/Italy	Car industry	Assembly work	Bodyworks		54 Robocarriers			
1979	Wasabroed Filipstadt/Sweden	Food industry	Servicing of production	Bread		17 Robotrailers			

Installations of Digitron, Switzerland

YEAR OF SALE	USER/SITE	BRANCH OF INDUSTRY	APPLICATION	TRANSPORT GOODS LOADING EQUIPMENT	Transport weight in tonnes	TRUCK TYPE/ TRUCK MAKE	NETWORK Length in m	NETWORK Number of switching stops	NETWORK Number of stopping points
1979	Deutsche Shell, Hamburg/Federal Republic of Germany	Chemical industry	Distribution and transport in the storage area	Lubricants		14 Robotrailers	950		
1979	Volvo BM, Eskilstuna/Sweden	Truck construction	Assembly work	Agricultural machinery		2 Robotrailers			
1979	Fiat Mirafiori, Turin/Italy	Car industry	Assembly work	Engines		37 Robotrailers			
1979	Citroen Lorraire, France	Car industry	Assembly work	Engines		255 Robomatics			
1980	Electrolux, Mariestad/ Sweden	Electrical industry		Household appliances		4 Robotrailers			

Installations of Digitron, Switzerland

APPENDIX 263

| YEAR OF SALE | USER/SITE | BRANCH OF INDUSTRY | APPLICATION | TRANSPORT GOODS LOADING EQUIPMENT | Transport weight in tonnes | TRUCK TYPE/ TRUCK MAKE | NETWORK |||
							Length in m	Number of switching stops	Number of switching stopping points
1981	Philips VCR-Werk, Vienna/Austria	Electrical industry		Video recorders		30 Robotrailers			
1981	Opel, Rüsselheim/Federal Republic of Germany	Car industry	Assembly work	Engines	0,3 t	100 Robomatic	600		
1981	Philips, Norrköping/Sweden	Electrical industry		Television sets		4 Robotrailers			
1981	ASEA, Ludvika/Sweden					4 Robotrailers			
1975	Volvo, Ghent Belgium	Automobile industry	Assembly	Car bodies		50 Low carrier			

Installations of Digitron, Switzerland

YEAR OF SALE	USER/SITE	BRANCH OF INDUSTRY	APPLICATION	TRANSPORT GOODS LOADING EQUIPMENT	Transport weight in tonnes	TRUCK TYPE/ TRUCK MAKE	NETWORK Length in m	Number of switching stops	Number of stopping points
1981	Sihlpost Zürich/Switzerland	Post office	Transport	Letters and parcels		3 Robotrailers			
1982	Télémechanique Nice/France	Electrical industry	Warehousing			1 Robotrailer			
1982	IBM-Järfälla/Sweden	Electrical industry	Transport	EDP parts		10 Robotrailer			
1982	Télémechanique Nice/France	Electrical industry	Assembly	EDP assembly and test		1 Robomatic			
1982	General Motors Aspern/Austria	Automobile industry	Assembly	Motors		72 Robomatic			

Installations of Digitron, Switzerland

APPENDIX

| YEAR OF SALE | USER/SITE | BRANCH OF INDUSTRY | APPLICATION | TRANSPORT GOODS LOADING EQUIPMENT | Transport weight in tonnes | TRUCK TYPE/ TRUCK MAKE | NETWORK |||
							Length in m	Number of switching stops	Number of stopping points
1982	Elektrolux Matiestad/Sweden	Electrical industry	Supply and retrieval in production			5 Robotrailer			
1982	Fasa-Renault Valladolid/Spain	Automobile industry	Supply and retrieval in production	Motors		21 Robomatic			
1983	Postverket Stockholm/Sweden					75 Robomatic			
1983	Mühlbach-Papier AG Lupfig/Switzerland	Paper industry	Warehousing	Pallets		12 Robotrailer			
1983	MBB Donauwörth/West Germany		Supply and retrieval in production	Tools		6 Robotrailer			

Installations of Digitron, Switzerland

YEAR OF SALE	USER/SITE	BRANCH OF INDUSTRY	APPLICATION	TRANSPORT GOODS LOADING EQUIPMENT	Transport weight in tonnes	TRUCK TYPE/ TRUCK MAKE	NETWORK		
							Length in m	Number of switching stops	Number of stopping points
1984	Opel Bochum/West Germany	Automobile industry	Assembly	Car doors		240 Robomatic			
1984	General Motors Antwerpen/Belgium	Automobile industry	Assembly	Car doors		240 Robomatic			

Installations of Digitron, Switzerland

APPENDIX

YEAR OF SALE	USER/SITE	BRANCH OF INDUSTRY	APPLICATION	TRANSPORT GOODS LOADING EQUIPMENT	Transport weight in tonnes	TRUCK TYPE/ TRUCK MAKE	NETWORK		
							Length in m	Number of switching stops	Number of stopping points
	A. C. Delco England								
	Alif Lataur France								
	ASDA England								
	Bridon Wire England								
	British Army England								

Installations of Barrett

YEAR OF SALE	USER/SITE	BRANCH OF INDUSTRY	APPLICATION	TRANSPORT GOODS LOADING EQUIPMENT	Transport weight in tonnes	TRUCK TYPE/ TRUCK MAKE	NETWORK Length in m	NETWORK Number of switching stops	NETWORK Number of stopping points
	British Leyland England								
	Coca Cola England								
	Green King Ltd. England								
	H. K. Heinz England								
	IBM England								

Installations of Barrett

APPENDIX

YEAR OF SALE	USER/SITE	BRANCH OF INDUSTRY	APPLICATION	TRANSPORT GOODS LOADING EQUIPMENT	Transport weight in tonnes	TRUCK TYPE/ TRUCK MAKE	NETWORK Length in m	Number of switching stops	Number of stopping points
	Imperial Chemical England								
	J. Lyons England								
	Linread Ltd. England								
	Migros Switzerland								
	Naoussa Spinning Greece								

Installations of Barrett

| YEAR OF SALE | USER/SITE | BRANCH OF INDUSTRY | APPLICATION | TRANSPORT GOODS LOADING EQUIPMENT | Transport weight in tonnes | TRUCK TYPE/ TRUCK MAKE | NETWORK |||
							Length in m	Number of switching stops	Number of switching stopping points
	Papier Fabrik Switzerland								
	Reckitt and Colman England								
	Rosedale Industries England								
	Technische Unie Holland								
	Standard Romper Cumberland, R. I.								

Installations of Barrett

APPENDIX

YEAR OF SALE	USER/SITE	BRANCH OF INDUSTRY	APPLICATION	TRANSPORT GOODS LOADING EQUIPMENT	Transport weight in tonnes	TRUCK TYPE/ TRUCK MAKE	NETWORK Length in m	Number of switching stops	Number of stopping points
1980	August Köhler Papierwerk Oberkirch	Paper industry		Paper rolls	17	1 Forktruck			
1980	Felix Eslov/Sweden	Food industry		Deep frozen food on pallets	0,5	1 Forktruck			
1980	Scan Vast Gothenburg/Sweden	Food industry		Pallets	0,5	1 Truck with roller-beds			
1981	General Motors Antwerpen/Belgium	Automobile industry		Motors	0,2	110 Trucks			
1981	Stora Kopparberg Stalldalen Stalldalen/Sweden	Paper industry		Paper rolls	3,0	1 Truck			

Installations of Carrago, Italy

YEAR OF SALE	USER/SITE	BRANCH OF INDUSTRY	APPLICATION	TRANSPORT GOODS LOADING EQUIPMENT	Transport weight in tonnes	TRUCK TYPE/ TRUCK MAKE	NETWORK		
							Length in m	Number of switching stops	Number of stopping points
1981	Fata Italy	Engineering		Pallets		1 Truck			
1981	Fata Italy	Engineering	Supply and retrieval in production	Single items		2 Forktrucks			
1981	Kimberly Clark Ltd Prudhoe/Great Britain	Paper industry		Paper rolls	2,2	6 Forktrucks			
1981	Citroen Tremery Tremery/France	Automobile industry		Motors	0,2	70 Trucks			
1981 1982	Leche Pascal Burgos/Spain	Food industry		Milk on pallets	1,0	3 Trucks 1 Truck (1982)			

Installations of Carrago, Italy

APPENDIX

YEAR OF SALE	USER/SITE	BRANCH OF INDUSTRY	APPLICATION	TRANSPORT GOODS LOADING EQUIPMENT	Transport weight in tonnes	TRUCK TYPE/ TRUCK MAKE	NETWORK Length in m	Number of switching stops	Number of switching stopping points
1981	Wulkan Anderstorp/Sweden	Engineering		Screws on pallets	1,5	1 Forktruck			
1982 1983	Hoechst AG Werk Kelheim	Chemical industry		Containers	5,0	1 Special forktruck 3 Special forktrucks (1983)			
1982	Elektrolux Mariestad/Sweden	Engineering		Pallets	1,0	3 Forktrucks			
1982	Svenska Tobaks AB Malmo/Sweden	Tobacco industry		Pallets	1,0	2 Forktrucks			
1982	Citroen Rennes/France	Automobile industry		Car body parts	0,5	12 Trucks			

Installations of Carrago, Italy

YEAR OF SALE	USER/SITE	BRANCH OF INDUSTRY	APPLICATION	TRANSPORT GOODS LOADING EQUIPMENT	Transport weight in tonnes	TRUCK TYPE/ TRUCK MAKE	NETWORK Length in m	Number of switching stops	Number of stopping points
1982	Citroen Meudon/France	Automobile industry		Machined parts		4 Trucks			

Installations of Carrago, Italy

APPENDIX

YEAR OF SALE	USER/SITE	BRANCH OF INDUSTRY	APPLICATION	TRANSPORT GOODS LOADING EQUIPMENT	Transport weight in tonnes	TRUCK TYPE/ TRUCK MAKE	NETWORK		
							Length in m	Number of switching stops	Number of stopping points
1979	Ersboda Dairy Sweden	Food industry		Milk in cartons on pallets	0,4	7			
1981	Modo Sweden	Paper industry		Cartons of paper on pallets	0,7	1			
1981	Ford Spain	Automobile industry	Supply and retrieval in production	Components in containers	1,0	1			
1982	P.S.F. Great Britain	Automobile industry	Supply and retrieval in production	Body parts	0,3	3			
1982	VAZ USSR	Automobile industry		Chassis parts in pallets	0,3	23			

Installations of FATA, Italy

YEAR OF SALE	USER/SITE	BRANCH OF INDUSTRY	APPLICATION	TRANSPORT GOODS LOADING EQUIPMENT	Trans- port weight in tonnes	TRUCK TYPE/ TRUCK MAKE	NETWORK		
							Length in m	Number of switching stops	Number of stopping points
1983	Snia Viscosa USSR	Textile industry		Acrylic fibres on pallets	0,6	13			
1983	Post Office Sweden			Pallets	0,8	60			

Installations of FATA, Italy

References

(1) Baumgarten; Böckmann; Gail — Voraussetzungen automatisierter Lager *(Requirements for automated stores)*; Beuth Verlag GmbH, Berlin, Köln 1978

(2) Bundesminister für Forschung und Technologie (editor) — Schriftenreihe "Humanisierung des Arbeitslebens" Band 3, Gruppenarbeit in der Motorenmontage *(Study series "Humanisation of working life" Volume 3, Group work in engine assembly)*; Campus Verlag, Frankfurt 1979

(3) Gudehus, T. — Transportsysteme für leichtes Stückgut *(Transport systems for light piece goods)*; VDI-Verlag, Düsseldorf 1977

(4) Hartmann, B. — Betriebswirtschaftliche Grundlagen der automatisierten Datenverarbeitung *(Industrial management principles of automated data processing)*; R. Haufe Verlag 4th Edition, Freiburg 1979

(5) Hartmann, B. — Betriebswirtschaftliche Grundlagen der automatisierten Datenverarbeitung *(Industrial management principles of automated data processing)*; R. Haufe Verlag First Edition, Freiburg 1961

(6) Heinen, E. — Industriebetriebslehre *(Industrial management)*; Th. Gabler Verlag 6th Edition, Wiesbaden 1978

(7) Maynard, H. B. — Handbook of Modern Manufacturing Management; McGraw Hill, New York 1970

(8) Mellerowicz, K. — Betriebswirtschaftslehre der Industrie *(Industrial management)*; Haufe Verlag 7th Edition Volume 1 Freiburg 1981

(9) Monsberger, J. — Kreisförderer und Schleppkreisförderer, in Salzer, G. (Hrsg.), Stetigförderer Teil 2 *(Circular conveyors and drag circular conveyors in Salzer, G. (Editor), Conveyor systems Part 2)*; Krausskopf Verlag, Mainz 1967

(10) Müller, Th.   Innerbetriebliche Transportsysteme – Anforderungskriterien und Einsatzmöglichkeiten, in Baumgarten, H. u.a. (Hrsg.) RKW-Handbuch Logistik *(Internal transport systems – Prerequisites and applications, in Baumgarten, H. (Editor) et al. RKW handbook Logistics)*; Erich Schmidt Verlag, Berlin 1981

(11) Salzer, G. (Hrsg.)   Stetigförderer Teil 2 *(Conveyor systems Part 2)*; Krausskopf Verlag, Mainz 1967

(12) Tietze, Schenk   Halbleiter Schaltungstechnik *(Semiconductor circuit technology)*; Springer Verlag 5th Edition, Berlin Heidelberg New York 1980

(13) Zangemeister, C.   Nutzwertanalysen in der Systemtechnik *(Economic value analysis in systems technology)*; Wittmannsche Buchhandlungen, Munich 1970

(14) Aguren, S. et al.   The Volvo Kolmar Plant; The Rationalisation Council SAF/Lo, Stockholm 1976

(15) Armbruster, R.   Schwachstellen bei der Inbetriebnahme aus der Sicht des Betreibers *(Inadequacies during commissioning from the user's viewpoint)*; VDI report 433, VDI-Verlag, Düsseldorf 1981

(16) Baier, W.   Hoffnung auf Natrium und Schwefel *(Hope for sodium and sulphur)*; Elektrische Energie-Technik, 1981 No. 7

(17) Baumgarten, H.   Fahrerlose Transportsysteme (FTS) zur Automatisierung des innerbetrieblichen Materialflusses *(Driverless transport systems (DTS) for the automation of the internal material flow)*; "Automation of internal material flow" Symposium, Karlovy Vary 1980

(18) Baumgarten, H.   Arbeitsunterlage "Flurförderzeuge" zur Vorlesung Fördertechnik (WS 80/81) *(Working documents "Industrial trucks" for the transport technology lectures (Winter term 80/81))*; Specialist area Transport and storage technology/logistics at the Technical University of Berlin

(19) Baumgarten, H.   Arbeitsunterlage "Schleppkreisförderer" zur Vorlesung Fördertechnik (WS 80/81) *(Working documents "Circular drag conveyors" for the transport technology lectures (Winter term 80/81))*; Specialist area Transport and storage technology/logistics at the Technical University of Berlin

REFERENCES

(20) Baumgarten, H.   Arbeitsunterlage "Einschienenhängebahn" zur Vorlesung Födertechnik (SS 82) *(Working documents "Monorail conveyors" for the transport technology lectures (Summer term 82))*; Specialist area Transport and storage technology/logistics at the Technical University of Berlin

(21) Baumgarten, H.   Field Analysis of Driverless Transport Systems in the FRG; 3rd International Conference on Automation in Warehousing, Chicago 1979

(22) Baumgarten, H.   Marktuntersuchung und Systemstudie über Fahrerlose Transportsysteme *(Market analysis and system study on driverless transport systems)*; unpublished study of the specialist area Transport and storage technology/logistics at the Technical University of Berlin 1979

(23) Beckert, R.   Alternative zum Montage-Fließband, selbstfahrende Werkstückträger *(Alternatives to the assembly line, self-driving workpiece carriers)*; Förden und Heben, 1980 No. 2

(24) Boehme; Müller   Betriebskostenvergleich innerbetrieblicher Transportsysteme *(Operating cost comparison of internal transport systems)*; unpublished study of the specialist area transport and storage technology/ logistics at the Technical University of Berlin 1981

(25) Boldrin, B.   The Great AGVS Race; Materials Handling Engineering, June 1980

(26) Engler, W.   Realisierung von schlüsselfertigen EDV-Systemen für Disposition und Steuerung von automatisierten Materialflußsystemen *(Implementation of turnkey computer systems for the organisation and control of automated material flow systems)*; VDI report 433, VDI-Verlag, Düsseldorf 1981

(27) Geisselreither, W.   Möglichkeiten der Programmierung *(Possibilities of programming)*; First Berman DTS Conference, Munich 1972

(28) Glätzel, F.   Bericht über das 5. Internationale Elektro-Fahrzeug-Symposium *(Report on the 5th International Electric Truck Symposium)*; Elektrotechnische Zeitschrift, 1979 No. 3

(29) Gunsser, P.   Steuerung fahrerloser Transportsysteme *(The control of driverless transport systems)*; Der Elektroniker, 1973 No. 10

(30) Gunsser, Krug — Erfahrungsbericht über den Einsatz von programmierbaren Steuerungen bei FTS *(Experience report on the use of programmable controllers with DTS)*; VDI report 327, VDI-Verlag, Düsseldorf 1978

(31) Gunsser, P. — Materialfluß-Steuerung im Lagervorhof *(Material flow control in the store forecourt)*; Fördern und Heben, Marktbild Lager 81/82, 1981

(32) Gunsser, P. — Zentrale und dezentrale Steuerung von fahrerlosen Flurförderzeugen *(Central and local control of driverless industrial trucks)*, Transmatic 76 Part 1, Krausskopf Verlag Mainz 1976

(33) Hauch, W. — Automatisch Fahren und Laden *(Automatic transport and loading)*; Information brochure PHB, Schwieberdingen 1980

(34) Hesser, P. — Schwachstellen bei Ausschreibung, Angebotsvergleich und Vergabe aus der Sicht des Betreibers *(Weak points in tenders, adjudication and award of contract from the user's viewpoint)*; VDI report 433, VDI Verlag Düsseldorf 1981

(35) Iseli, I. — Einfluß des Steuerungskonzeptes auf die Zuverlässigkeit von Hochregallagern *(Effect of the control design philosophy on the reliability of high-bay warehouses)*; Technical communications from Sprecher und Schuh AG, Aarau Switzerland 1978

(36) Klug, H. G. — Entwicklungstendenzen von Fahrerlosen Transportsystemen unter Berücksichtigung des Mikroprozessor- und Softwareeinsatzes, in Baumgarten, H. (Hrsg.) Logistik im Unternehmen "Berliner Symposium '79" *(Development trends for driverless transport systems with consideration of the application of microprocessors and software, in Baumgarten, H. (Editor) Logistics in the factory "Berliner Symposium 79")*; Krausskopf Verlag, Mainz 1980

(37) Klug, H. G. — Projektierung von FTS-Anlagen *(Planning of DTS installations)*; MIC Congress Munich 1980

(38) Marx; Wagner — Stapler sind zum Stapeln da *(Stackers are there for stacking)*; Materialfluss 1972, No. 7

(39) Müller, Th. — Stand der Entwicklung von Fahrzeugsystemen unter Berücksichtigung der Fahrerlosen Transportsysteme, in Baumgarten, H. (Hrsg.) Logistik im Unternehmen "Berliner Symposium '79" *(State of development of truck systems with consideration of driverless transport systems, in Baumgarten, H.*

REFERENCES

	(Editor) Logistics in the factory "Berliner Symposium 79"); Krausskopf Verlag, Mainz 1980
(40) Müller, Th.	Der wirtschaftliche Einsatzbereich von Hochregalstaplern bei der Ver- und Ent-sorgung von Palettenregalen *(The economic range of applications of stacker cranes in the servicing of pallet racking)*; Fördern und Heben, 1982 No. 3
(41) Müller; Bünsow	Möglichkeiten der Bewertung verschiedener Transportarten *(Possibilities of evaluating different forms of transport)*; Fördern und Heben, 1982 No. 9
(42) no author	Märkte, Tendenzen und Entwicklungen der Fördertechnik *(Markets, trends and developments in transport technology)*; Fördern und Heben, 1981 No. 1
(43) no author	Barrett Bulletin A2, Informationsprospekt der Fa. Barrett *(Information prospectus Barrett)*; Northbrook Illinois USA no year specified
(44) no author	Über passivem Leitband *(Via the passive guide path)*; Materialfluss, 1967 No. 7
(45) no author	psb (Pfalz-Stahlbau) gesammelte Firmenprospekte und Referenzlisten *(psb (Pfalz-Stahlbau) collected company prospectuses and reference lists)*; Pirmasens 1979
(46) no author	Teletrak-Systemordner *(Teletrak system collator)*; technical communications of Jungheinrich, Hamburg-Norderstedt 1980
(47) no author	Komatsu News; Communications from Komatsu, Tokyo Japan, 1976, No. 2
(48) no author	Entwurf zur VDI-Richtlinie 3663 *(Draft for VDI recommendation 3663)*; VDI handbook Material flow and transport technology VDI-Verlag, Düsseldorf 1982
(49) no author	Schindler-Digitron Presse-Information Nr. 03-15/126 *(Schindler-Digitron Press release No. 03-15/126)*; Brügg-Biel Switzerland 1982
(50) no author	Gabelstapler eingebunden in EDV-gestützte Lagerhaltung *(Fork-lift trucks engaged in computer-based store control)*; Produktion, 1982 No. 35
(51) no author	Induktionsschleifen-gesteuerte Flurfördermittel *(Induction loop controlled industrial transport systems)*; Förden und Heben, 1982 No. 8

(52) v. Revenstorff, C. Die FTS-Technik mausert sich *(DTS technology is moulting)*; Materialfluss, 1977 No. 4

(53) Reuff, H. Beschreibung der Ablauforganisation im Pflichtenheft *(Description of the process organisation in the performance specification)*; VDI-report 433, VDI-Verlag Düsseldorf 1981

(54) Schick, I. Die Lenkregelung von Flurförderzeugen im Leitwegsystem *(Steering control of industrial trucks using the guide path system)*; Elektronik, 1969 No. 12

(55) Wenzel, R. Organisatorische Möglichkeiten von Arbeitsverteilsystemen *(Organisational possibilities of work distribution systems)*; Werkstattstechnik 1973, No. 5

(56) Werner, R. Schwachstellen bei der Ausschreibung und Vergabe aus der Sicht des Lieferanten *(Weak points in tenders and award of contract from the supplier's viewpoint)*; VDI report 433, VDI-Verlag Düsseldorf 1981

General Reading

Automated Guided Vehicle Systems, 1st International Conference, June 1981. Published by IFS (Publications), Bedford, UK, 250 pp.

Automated Guided Vehicle Systems, 2nd International Conference, June 1983, Published by IFS (Publications) Bedford, UK, 345 pp.

List of Illustrations and Tables

Figure	Title	Source
Fig. 1	Classification of continuous transporters for piece goods	Author
Fig. 2	Classification of internal truck systems for the transport of piece parts	Author
Fig. 3	Specifications for internal transport systems	Author
Fig. 4	System overview of the main characteristics of internal transport systems	Author
Fig. 5	System overview of the main characteristics of internal transport systems	Author
Fig. 6	Driver-operated industrial trucks with electric motor drive (battery). (a) Reach mast truck with driver and three-wheel construction. (b) Pallet truck with driver. (c) Truck tractor with four-wheel construction	Author
Fig. 7	Components of a roller conveyor system including belt conveyor for overcoming gradients	Rheinstahl
Fig. 8	Components of the circular traction conveyor (power and free). (a) Traction unit beneath the traction track. (b) Traction unit inside the traction track	/9/
Fig. 9	Components of the monorail conveyor	Rheinstahl
Fig. 10	Application areas of electrically powered rail conveyors and power and free conveyors	/20/
Fig. 11	Handling systems for servicing work places: (a) Container conveyor systems in	

Figure	Title	Source
	conjunction with a stacker crane operated live store; (b) Stacker crane system for servicing the work places and the pigeon-hole store (dual operation)	/45/
Fig. 12	Overview of the most important system components of driverless transport systems	Author
Fig. 13	Steering systems for industrial trucks, swivelling bolster steering and axle pivot steering systems	Author
Fig. 14	Envelopes for a truck with fifth-wheel steering	/46/
Fig. 15	Load transfer between a tractor train with lifting platform trailers and a stationary chain conveyor	Jungheinrich
Fig. 16	Load transfer from a tractor train to a conveyor by means of telescopic forks	Jungheinrich
Fig. 17	Loading of a tractor train inside a palletiser	Wagner
Fig. 18	Sequential unloading at the interface to a high-bay racking store with subsequent contour testing. Truck and station are equipped with chain conveyors	Wagner
Fig. 19	Simultaneous load transfer between stationary roller conveyors and trailers with roller conveyor segments mounted on them	Wagner
Fig. 20	Reversing tractor with bolt coupling	Wagner
Fig. 21	Reversing tractor with electromagnetic coupling for automatic coupling and decoupling operations	Jungheinrich
Fig. 22	Tractor train with true-tracking pallet trucks as trailers	Jungheinrich
Fig. 23	Semi-trailer truck with roller containers in commissioner applications for wholesale operations	Jungheinrich
Fig. 24	Inductively steered pallet truck; drawbar towing	Jungheinrich
Fig. 25	Inductively steered pallet truck with fork length for 2 pallets and infra-red distance control, cf chapter 3.5.3.2.	Wagner

LIST OF ILLUSTRATIONS AND TABLES

Figure	Title	Source
Fig. 26	Inductively steered pallet truck with upward folding platform	Wagner
Fig. 27	Unloading of load units (forwards travel) using pallet trucks	/46/
Fig. 28	Load handling (unloading and/or picking up) with automatically reversing pallet trucks	/46/
Fig. 29	Inductively steered pallet trucks in commissioner applications with lifting equipment	Wagner
Fig. 30	Skid tractor with differential steering	Schindler-Digitron
Fig. 31	Skid tractor transporting a motor including the loading unit	Schindler-Digitron
Fig. 32	Skid tractor transporting a raised roller container	Schindler-Digitron
Fig. 33	Steering systems of industrial trucks, differential steering and all-wheel steering	Author
Fig. 34	Steering systems of industrial trucks, transverse travel and diagonal travel with all-wheel steering	Author
Fig. 35	Skid tractor as a transport platform in production integrated operation (welding line)	Schindler-Digitron
Fig. 36	Truck with heavy link conveyor for transporting lorry engines	Wagner
Fig. 37	Truck with roller conveyor for handling loads during commissioning	ACS-Volvo
Fig. 38	Reach mast stacker in the side-loading version	Jungheinrich
Fig. 39	Reach mast stackers transporting 3 roller containers and handling loads in a cold store with two rack levels	ACS-Volvo
Fig. 40	Servicing of production work places with the help of an inductively steered telescopic fork lift truck	Wagner
Fig. 41	Special truck used in the final assembly of car engines Dead weight 300kg Payload 300kg	Schindler-Digitron

Figure	Title	Source
Fig. 42	Special truck in the aggregate assembly of chassis, car industry	Schindler-Digitron
Fig. 43	Determining factors for load transfer with driverless transport systems	Author
Fig. 44	Systematic classification of load transfer for driverless transport systems	Author
Fig. 45	Data flow for driverless transport systems	Author
Fig. 46	Classification of the data transfer system	Author
Fig. 47	Operating unit and display	/46/
Fig. 48	Magnetic field around a guide wire and its scanning	/29/
Fig. 49	Inductive steering principle	/46/
Fig. 50	Block diagram of the low-frequency generator for supplying the guide wire	/29/
Fig. 51	Pulse control for a DC series wound motor	/29/
Fig. 52	Example of a hierarchic control design philosophy	VDI-2690
Fig. 53	Factors influencing the control design philosophy of driverless transport systems	Author
Fig. 54	Example of a route display panel	/46/
Fig. 55	Options for on-board route control	Author
Fig. 56	Example of a decentralised stationary control system with central transport process computer	/32/
Fig. 57	Flow diagram for route control	Author/46/
Fig. 58	Coding example of a travel route with incremental position recognition	/27/
Fig. 59	Coding example for a network with absolute digital position recognition	/27/
Fig. 60	Cross section through a guide wire	/46/
Fig. 61	Example of a complete network installation using the single frequency principle	/46/
Fig. 62	Single blocking using the single frequency principle	/46/

LIST OF ILLUSTRATIONS AND TABLES

Figure	Title	Source
Fig. 63	Blocking of a confluence using the single frequency principle	/46/
Fig. 64	Dynamic and static cases for block control	/36/
Fig. 65	Block control using the multi-frequency principle	Author/Wagner
Fig. 66	Flexible block control using the multi-frequency principle	Author/Wagner
Fig. 67	Points switching using the single frequency principle	/46/
Fig. 68	Turning off from main routes using the multi-frequency principle controlled from the truck	Wagner
Fig. 69	Lift journeys with through travel	Wagner
Fig. 70	Internal structure of the MC 6800 microprocessor	/12/
Fig. 71	Block diagram of a microcomputer	/12/
Fig. 72	Layout of the network in an enterprise of the electrical equipment industry	Author/Siemens
Fig. 73	Layout of the network for the engine assembly application	Author/Wagner
Fig. 74	Final assembly of an engine with the help of AGVS trucks	Wagner
Fig. 75	System elements for the link-up of production, forecourt and storage areas (in accordance with VDI recommendation 2690)	VDI 2690
Fig. 76	Diagram of the commissioning system and the computer hierarchy in a publishing distribution centre	Author/VVA
Fig. 77	Examples of the structure of data telegrams between conveyor control, the AGVS and the central process computer	VVA
Fig. 78	Layout of the network in a commissioning system in the wholesale trade	Author
Fig. 79	Assessment of the viability of AGVS by the users of all the systems studied	/22/
Fig. 80	Operating period (relative frequency) of the studied systems	/22/

Figure	Title	Source
Fig. 81	Battery charging procedure (relative frequency) in the systems studied	/22/
Fig. 82	Assessment of experience with AGVS by users of operational and shut-down systems	/22/
Fig. 83	Planning difficulties for AGVS from the user's viewpoint (arithmetic mean of the given evaluation grades 1–10 1 = no problem, 10 = most difficult)	/22/
Fig. 84	Reasons for not introducing AGVS (relative frequency) in past projects	/22/
Fig. 85	Sankey diagram of a material flow analysis (Amounts in load units/time unit)	Wagner
Fig. 86	Combined matrix for representing the quantity flows in load units (LE) per time unit and the associated transport distances (m)	Author
Fig. 87	Flow diagram for carrying out a simulation study	/48/
Fig. 88	Investment calculation procedures	/6/
Fig. 89	Operating costs (DM/LU) as a function of the Transport Quantity (LU/H) for all systems. Transport Distance is 25M and LU is Load Unit	Author/24/
Fig. 90	Operating Costs (DM/LU) as a function of the Transport Quantity for all systems. Transport Distance is 50M	Author/24/
Fig. 91	Operating Costs (DM/LU) as a function of the Transport Quantity for all systems. Transport Distance is 100M	Author/24/
Fig. 92	Operating Costs (DM/LU) as a function of the Transport Quantity (LU/H) for all systems. Transport Distance is 200M and LU is Load Unit	Author/24/
Fig. 93	Operating Costs (DM/LU) as a function of the Transport Quantity (LU/H) for all systems. Transport Distance is 400M and LU is Load Unit	Author/24/
Fig. 94	Operating Costs (DM/LU) as a function of	

Figure	Title	Source
	the Transport Quantity for all systems. Transport Distance is 800M	Author/24/
Fig. 95	Operating Costs (DM/LU) as a function of the Transport Quantity for all systems. Transport Distance is 1600M	Author/24/
Fig. 96	Operating Costs (DM/LU) as a function of the Transport Quantity for all systems. Transport Distance is 3200M and LU is Load Unit	Author/24/
Fig. 97	Operating Costs (DM/LU) as a function of the Transport Distance (M) for all systems. Transport Quantity is 4 (LU/H)	Author/24/
Fig. 98	Operating Costs (DM/LU) as a function of the Transport Distance (M) for all systems. Transport Quantity is 8 (LU/H)	Author/24/
Fig. 99	Operating Costs (DM/LU) as a function of the Transport Distance (M) for all systems. Transport Quantity is 16 (LU/H)	Author/24/
Fig. 100	Operating Costs (DM/LU) as a function of the Transport Distance (M) for all systems. Transport Quantity is 32 (LU/H)	Author/24/
Fig. 101	Operating Costs (DM/LU) as a function of the Transport Distance (M) for all systems. Transport Quantity is 64 (LU/H)	Author/24/
Fig. 102	Operating Costs (DM/LU) as a function of the Transport Distance (M) for all systems. Transport Quantity is 128 (LU/H)	Author/24/
Fig. 103	Operating Costs (DM/LU) as a function of the Transport Distance (M) for all systems. Transport Quantity is 256 (LU/H)	Author/24/
Fig. 104	Operating Costs (DM/LU) as a function of the Transport Distance (M) for all systems. Transport Quantity is 512 (LU/H)	Author/24/
Fig. 105	Reasons for implementing AGVS from the user's viewpoint (arithmetic mean of the given evaluation grades 1-10; 1=most unimportant; 10=most important)	/22/

£28
SCC

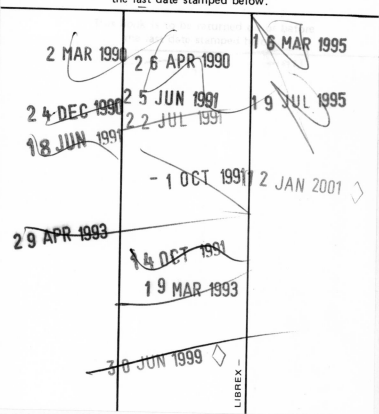